Financial Feasibility Studies for Property Development

Essential for any real estate professional or student performing feasibility studies for property development using Microsoft Excel and two of the most commonly used proprietary software systems, Argus Developer and Estate Master DF.

This is the first book to not only review the place of financial feasibility studies in the property development process, but to examine both the theory and mechanics of feasibility studies through the construction of user friendly examples using these software systems. The development process has seen considerable changes in practice in recent years as developers and advisors have adopted modern spreadsheets and software models to carry out feasibility studies and appraisals. This has greatly extended their ability to model more complex developments and more sophisticated funding arrangements, saving time and improving accuracy.

Tim Havard brings over 25 years of industry and software experience to guide students and practitioners through the theory of development appraisals and feasibility studies before providing internationally applicable worked examples and potential pitfalls using Excel, Argus Developer and Estate Master DF.

Dr Tim Havard is the author of two books, *Investment Property Valuation Today* and *Contemporary Property Development*, now in its 2nd edition. He has been in the property industry for more than 25 years as a surveyor and academic.

Financial Feasibility Studies for Property Development

Theory and Practice

Tim Havard

LONDON AND NEW YORK

First published 2014
by Routledge
2 Park Square, Milton Park, Abingdon, Oxon OX14 4RN

Simultaneously published in the USA and Canada
by Routledge
711 Third Avenue, New York, NY 10017

Routledge is an imprint of the Taylor & Francis Group, an informa business

British Library Cataloguing in Publication Data
A catalogue record for this book is available from the British Library

Library of Congress Cataloging in Publication Data
Havard, Timothy.
  Financial feasibility studies for property development: theory and
  practice/Tim Havard.
  pages cm
  Includes bibliographical references and index.
  1. Real estate development – Finance. 2. Real property – Valuation.
  3. Real estate development. 4. Real estate development – Finance –
  Case studies. 5. Real property – Valuation – Case studies. 6. Real
  estate development – Case studies. I. Title.
  HD1390.H37 2014
  333.33 – dc23
  2013010376

ISBN: 978-0-415-65916-1 (hbk)
ISBN: 978-0-415-65917-8 (pbk)
ISBN: 978-0-203-64022-7 (ebk)

Typeset in Bembo
by Florence Production Ltd, Stoodleigh, Devon, UK

Contents

List of illustrations   xi

1 Introduction   1

Scope and purpose of the book   1
The structure of this book   1

PART ONE
Principles of development modelling   3

2 The background to development appraisal   5

Definition of property development   5
The property markets and development   6
Property developers   6
Changes in the development environment   9

3 The theory of development financial feasibility studies   12

*Purposes of development appraisal and its role in the development
     process   12*

4 The basic development appraisal equation   17

5 Development preliminaries and initial project
considerations   22

The property development process in outline   22
Establishing project objectives   22
Different types of development objectives   22
Social objectives   23
Financial objectives   24

**6 Establishing development constraints – the development
envelope**   **25**

Planning   26
Contractual   26
Political   28
Environmental   28
Market timing   29
Financing development schemes   30
Equity funds   30
Debt finance   30
Senior debt loan   32
Mezzanine loan   32
Joint venture 100 per cent finance   32
Grants and public sources of finance   32
*Different procurement methods and their influence on the development
    appraisal   33*

**7 The financial appraisal of development schemes –
investigations required to construct the appraisal**   **35**

Site and land assembly   35
Data collection required for a development appraisal   36
Planning data required   36
Location   37
Topography, vegetation, heritage, etc.   38
Market analysis and appraisal   41
Market trends and analysis   41
Sales and leasing supply and demand   43
Financial and economic trends   44
Determining building dimensions   45
*Determining the timing of the development and its stages and
    phases   45*

Estimating development costs   48

    Build costs   48
    Construction costs   48

*Estimating approaches – superficial area, elemental and detailed
    breakdown   49*
Sources of cost information   50
The cost of the professional team   50

    The construction professionals   50

Estimating other development costs — marketing, disposal, statutory costs,
road works etc. *51*

    Contributions   51
    Land holding costs   52
    Selling and other costs   53
    Legal fees   53
    Other costs   53
    Dealing with intangible or uncertain costs   54

Estimating development values   *54*
Residential projects or elements   *54*
Commercial property — non-specialised   *54*
Rental value   *55*
Lease length   *56*
Lease terms   *56*
Physical package   *56*

    Physical package — size   56
    Physical package — shape   57
    Physical package — specification   57
    Physical package — condition   57

Location   *57*
    Macro location   57
    Micro location   58
Tenant status/size   *58*
Market conditions (boom/recession)   *58*
Yield and capital value   *58*
Sub variables that influence yields   *59*

    Tenant status   59
    Lease length   59
    Lease terms: break clauses   60
    Lease — other significant terms   60
    Physical properties — nature and quality of the building   60
    Location   60

Commercial property — specialised   *60*
Conclusions   *61*

8 The mechanics of constructing development
financial feasibility studies   **62**
An overview of the development of practice — from tables to proprietary
    software systems   *62*
Moving the model on: cash flows and spreadsheets   *68*

Residual (accumulative) cash flows and discounted cash flows 69
The residual (accumulative) cash flow calculation 75
The discounted cash flow calculation 76
Conclusions 78

9 **Sensitivity analysis – risk and uncertainty in
   development projects** 80

10 **Conclusions to Part One** 90

PART TWO
Development appraisal in practice 91

11 **Development appraisal using MS Excel** 93

Vulnerability of spreadsheet types to typical sources of error 106
Conclusions on the use of Excel in development appraisal 108

12 **Modelling development financial feasibility in Argus
    Developer – software introduction and case studies** 110

An overview of Argus Developer 110

History 110
An introduction to the programme 110
The project tab 111
The definition tab 113
The cash flow tab 116
The summary tab 118

Other features 119
Case studies – commercial development appraisal 121
*Argus Developer case study one – simple one building/one type/
one phase commercial project 121*

Land bid calculation 122

Argus Developer case study two – profitability calculation 149
More complex projects 152

Introduction 152

Argus Developer case study three – multiple buildings/types 152
Argus Developer case study four – mixed-use buildings 160
Argus Developer case study five – multiple timings/single phase 167
Argus Developer case study six – multiple phasing 172

Argus Developer case study seven – appraisal using multiple interest sets 178

Argus Developer case study eight – specialist properties: operated assets – golf course development 185

Modelling Residential Development Appraisal 196

Argus Developer case study nine – simple residential project: single building/type/single phase project 197

Land bid calculation 197

Argus Developer case study ten – more complex residential projects 204

Argus Developer case study eleven – modelling risk in Argus Developer 213

The sensitivity analysis module in Argus Developer 213

13 Modelling development financial feasibility in Estate Master DF – software outline and case studies **219**

An overview of Estate Master DF 219

Estate Master DF case study one – profitability calculation 229

Estate Master DF case study two – sensitivity analysis 239

Estate Master DF case study three – modelling a small mixed use project using Estate Master DF Lite 246

Introduction to Estate Master DF Lite 246

Case study model 246

Conclusion 259

14 Conclusions **260**

Appendix A – The Town and Country Planning (Use Classes) Order 261

Index 266

List of illustrations

2.1   The 20 leading house builders in the UK. This table is copyright
       The Construction Index and is reproduced with permission
2.2   Housing associations in the UK in 2011 (source:
       insidehousing.co.uk) PERMISSION PENDING
4.1   Development appraisal involves several steps. The first is to assess
       what the costs will be to develop the site (this will include the
       purchase of the land, the site preparation, the building costs and
       professional fees etc.) and the receipts from the sale or letting of the
       buildings produced. The second thing is to determine how long the
       process will take and when the expenditure and receipts will take
       place. This process is presented in this diagram, which represents a
       simple development with time running from left to right. This
       development starts with the major expenditure of purchasing the
       land. Monies are then spent on developing the site and then finally
       these buildings are sold. Although this is a simple project it is
       representative of the tone of cash flows in many projects.
4.2   In the appraisal itself, the next step is to total up all the costs and the
       receipts. This accumulative process is typical of most development
       appraisal models.
4.3   The next step in the appraisal process is to account for the effect of
       time on the project viability. In most cases this is done by
       calculating the interest charge that will be accrued on the
       development expenditure. In practice this interest may have
       different values on different elements of the project (land,
       construction etc.) and from different sources (equity, principle debt,
       mezzanine finance etc.). To simplify the calculation, many appraisals
       are done with an assumption of 100% debt financing and with an
       equal cost (see the comments on opportunity cost in text). In
       addition the traditional residual approach to appraisal makes a gross
       simplification in interest cost calculation (see later section). One of
       the key advantages that both the cash flow and the proprietary
       models have over the traditional residual is that they calculate
       interest more accurately (subject to the underlying cash flow being
       forecast accurately). The surplus at the end of the project is now net
       of interest.

4.4   The final step in most appraisals, true for both land residuals and
        profit calculations, is to discount the future surplus back to the
        present. In both cases the surplus is a future value, only available in
        the future. Given the long nature of development projects this can
        be several months or even several years in the future and as £1 paid
        at any time in the future will always be worth less than £1 today,
        discounting is required. Normally this discounting is done using the
        finance rate of interest
7.1   Typical expenditure pattern in a development project
8.1   Calculation of value on completion
8.2   Calculation of construction costs, professional fees and finance
        charges
8.3   Calculation of ancillary costs and present value of land
8.4   Base project development assumptions
8.5   Pre-construction phase of cash flow
8.6   Construction phase of cash flow
8.7   Post-construction phase of project
8.8   Performance analysis
8.9   Principles involved in traditional residual and residual (accumulative)
        cash flow approaches
8.10   Extract from cash flow: land purchase and pre-construction/
        planning stage
8.11   Extract from cash flow: construction stage
8.12   Extract from cash flow: letting-up and sales stage
8.13   Extract from cash flow showing formulas
8.14   The calculation principles involved in a DCF cash flow calculation.
        Each individual cash flow, both costs and receipts, are discount back
        to the present, without the calculation of interest. The discount rate
        used is normally the cost of finance
8.15   Extract from cash flow: land purchase and pre-construction/
        planning stage
8.16   Extract from cash flow: construction stage
8.17   Extract from cash flow: letting-up and sales stage
8.18   Extract from cash flow showing formulas
8.19   Project IRR calculation
9.1   Base assumption results
9.2   Rental values reduced by 10%
9.3   Results of rental values reduced by 10%
9.4   Capitalisation yields increased by 10%
9.5   Results of capitalisation yields increased by 10%
9.6   Required reduction in retail warehouse rents to extinguish
        development profitability
9.7   Development profitability extinguished
9.8   Inputs for a scenario modelling a market downturn
9.9   Results of scenario

9.10   Using Excel's scenario manager to define variables
9.11   Using Excel's scenario manager to model multiple development
        scenarios
9.12   Using Excel's scenario manager to model multiple development
        scenarios: results table
9.13   Probability extension to scenario modelling
11.1   Calculation of value on completion
11.2   Development assumptions used in calculation
11.3   Pre-construction phase of cash flow
11.4   Construction phase of cash flow
11.5   Post-construction phase of project
11.6   Performance analysis
11.7   Extract from the type one spreadsheet model showing formulas
11.8   Extract from Peter Brown's Development Appraisal Model
11.9   Spreadsheet produced by the Peter Brown model including example
        of formulas and logical time references
11.10   The initial screen of the HCA's DAT
11.11   HCA DAT site details screen
11.12   Residential property input screen
11.13   HCA DAT time framework entry
11.14   HCA DAT cost data entry
11.15   HCA DAT Non-residential property data entry screen
11.16   HCA DAT summary results
11.17   HCA DAT cash flow output extract
12.1   Argus Developer's initial screen
12.2   The assumptions for calculations
12.3   The timescale and phasing screen
12.4   The definition tab
12.5   Detail of definition tab – the grouped project type shortcuts
12.6   Drilling down behind to reveal the capitalisation of income screens
12.7   Detail of definition tab – construction cost and related fields
12.8   Drilling down behind the demolition section
12.9   Detail of demolition screen showing degree of variability/control
        possible with each element
12.10   Adjusting the timing and distribution of an expenditure
12.11   The cash flow tab – project view
12.12   The cash flow tab – adjustments
12.13   Accessing data distribution in the cash flow tab
12.14   The summary tab
12.15   The reporting module – note export options
12.16   Graphical analysis
12.17   Sensitivity analysis module
12.18   Project screen
12.19   Saving the project in Argus Developer
12.20   Expenditure assumptions

12.21   Receipts standard assumptions
12.22   Finance standard assumptions
12.23   Inflation and growth standard assumptions
12.24   Distribution standard assumptions
12.25   Calculation standard assumptions
12.26   Setting up the interest sets for the development
12.27   The standard residual tab setup
12.28   The residual screen once land residual only selected
12.29   The alternative targets for profit
12.30   The residual screen final setup
12.31   Data check warning screen which can be ignored at this stage
12.32   Extract from the default cash flow template showing associations
        and timings
12.33   The timescale and phasing shortcut
12.34   Timescale and phasing screen
12.35   The timescale and phasing screen as set up for our development
12.36   The definitions tab for our project
12.37   The shortcuts to the area screens
12.38   The capitalised rent area screen
12.39   Use type and areas
12.40   Construction calculation frame
12.41   Construction cost distribution
12.42   Rental calculation frame
12.43   The capitalisation of income frame
12.44   The data distribution screen for the disposal of the freehold interest
12.45   The completed area tab for our development
12.46   Definitions tab after the area screen has been completed
12.47   Detail of the hard landscaping assumptions override
12.48   The completed definitions tab
12.49   Overriding the related items setup
12.50   The cash flow tab
12.51   Adjusting the development model via the cash flow
12.52   Adjusting a cash flow line via the data distribution
12.53   Adjusting a cash flow line manually
12.54   The finance cash flow
12.55   Assumptions for calculation set-up for profit calculation
12.56   Land value entered into definition tab
12.57   Project tab for multi-building scheme
12.58   Timescales for the multi-building project
12.59   Office Type A area record
12.60   Office Type B area record
12.61   Retail Warehouse Type 1 area record
12.62   Retail Warehouse Type 2 area record
12.63   The completed definition tab for the multi-building scheme
12.64   Cash flow for the multi-building scheme

12.65   Project tab for the mixed-use development example
12.66   Definition tab for mixed use scheme
12.67   ITZA calculation screen
12.68   The completed commercial area screen for the retail part of the development
12.69   Unit sales area for the residential part of the scheme
12.70   The timing for the mixed use development with overlapping timings for the sales period
12.71   Overridden distribution of residential sales
12.72   Commercial sale timing moved to the end of the sale stage
12.73   Cash flow extract from mixed-use scheme
12.74   Types of building in development
12.75   The timing screen for the multiple building/multiple timing project
12.76   The original area screen for our project with base timings
12.77   Timing screen for the construction of the type 1 offices
12.78   Alteration of the assumed date of letting on the Lease tab
12.79   Adjusted construction cash flow (quarterly view)
12.80   Adjusted cash flow for the rental income streams
12.81   Project screen for multiple phase development
12.82   Timing screen for first phase
12.83   Phase 2 timing screen
12.84   Results summary ribbon illustrating the creation of three tabs when a two phase project is created. Note that the warehousing is the active phase
12.85   Phase 1 definition tab
12.86   Phase 2 definition tab
12.87   Merged phases tab
12.88   Merged cash flow for the uncorrected scheme, concentrating on the timing of land acquisition (semi-annual view cycle)
12.89   Corrected timescale
12.90   Merged phases cash flow of the corrected project
12.91   Multiple interest rates to model market interest rates movements
12.92   Resultant cash flow (semi-annual view and with results bar in alternative position)
12.93   Main construction loan
12.94   Land loan
12.95   Mezzanine loan
12.96   Area screen
12.97   Changing the interest set in the financial tab
12.98   Financial detail of demolition item phase 2
12.99   Merged phases cash flow with the row properties dialogue box highlighted
12.100 Row properties dialogue box
12.101 Interest button dialogue box
12.102 Changing the interest set

12.103 Appraisal of 'pay and play' golf course: project screen
12.104 Timescale and phasing set up for golf course development appraisal
12.105 Definition screen for golf course development
12.106 Details of architect's fees
12.107 The area screen with an operated asset selected
12.108 The operated assets editor with the two profiles
12.109 Number of rounds sold
12.110 Membership fees
12.111 Membership renewals and new sign ups per month
12.112 Operating revenues and expenses for golf course
12.113 Expenditure
12.114 Forecast bar and food receipts
12.115 Professional shop rent
12.116 Revenue stream calculation for clubhouse
12.117 Expenses calculation for clubhouse
12.118 Calculation of the value of clubhouse element
12.119 Construction of access roads for golf course
12.120 Construction of service buildings for golf course
12.121 Cash flow for golf course development
12.122 Developer opening screen
12.123 The residual assumptions
12.124 Interest sets assumptions
12.125 Project timescale assumptions
12.126 The single unit sales area
12.127 Sales construction timing
12.128 Definition tab
12.129 Cash flow
12.130 Project screen for more complex project
12.131 Stage one timescale
12.132 Stage two timescale
12.133 Single unit area sales sheet
12.134 The construction cost data entry area showing the two additional
        boxes at the bottom of the list
12.135 Construction cost distribution generated
12.136 Sales structure assumptions
12.137 Sales distribution for Phases 1 and 2 which results from the
        assumptions made within the single unit sales area screen
12.138 Phase 1 cash flow
12.139 Phase 2 cash flow
12.140 Merged phases cash flow
12.141 Simple commercial project
12.142 Accessing the sensitivity analysis module in Argus Developer
12.143 Initial sensitivity screen
12.144 Selection of construction element and the component (cost per m^2)
        to be tested in the sensitivity analysis

12.145 Setting the construction sensitivity parameters
12.146 Rent sensitivity and options
12.147 Rent sensitivity set up screen
12.148 Yield sensitivity set up screen
12.149 Sensitivity analysis outcome – rent and construction costs varied – yield static
12.150 Sensitivity analysis outcome – varying the yield using the slider function
13.1   Estate Master DF opening screen
13.2   Preference screen
13.3   Main inputs – preliminary
13.4   Land purchase
13.5   Land purchase – right-hand side of screen
13.6   Modelling professional fees associated with the development
13.7   Construction costs calculation sheet
13.8   Statutory fees
13.9   Tenancy schedule
13.10   Estate Master DF cash flow extract
13.11   Summary screen
13.12   Summary screen – performance indicators
13.13   Summary screen – financial analysis
13.14   Estate Master DF project input screen – simple commercial project
13.15   DF input screen – top
13.16   Land purchase and acquisition costs
13.17   The totals of the land acquisition and the associated costs
13.18   Definition and distribution of professional fees
13.19   The totalled sums of the professional fees
13.20   Construction costs entry and distribution
13.21   Construction costs entry and distribution totals
13.22   Miscellaneous costs – marketing
13.23   Financing assumptions
13.24   Finance screen totals
13.25   Tenancy schedule worksheet
13.26   Setting the lease details in the tenancy schedule
13.27   Setting the capitalisation rate and disposal value
13.28   DF cash flow extract
13.29   Summary sheet for simple commercial project
13.30   Performance measures
13.31   Performance measures – return on funds invested
13.32   Simple sensitivity table output
13.33   Probability-based scenario analysis set-up screen
13.34   Simulation output for development return
13.35   Simulation output for IRR
13.36   Estate Master DF Lite opening screen
13.37   Ribbon bar

13.38   Quick set-up screen 1
13.39   Quick set-up screen 2
13.40   Quick set-up screen 3
13.41   Quick set-up screen 4
13.42   Quick set-up screen 5
13.43   Quick set-up screen 6
13.44   Quick set-up screen 7
13.45   Quick set-up screen 8
13.46   Quick set-up screen 9
13.47   Quick set-up screen 10
13.48   Quick set-up screen 11
13.49   Quick set-up screen 11 – completed
13.50   Quick set-up screen 12
13.51   Quick set-up screen 13
13.52   Quick set-up screen 14
13.53   Quick set-up screen 15
13.54   Completed project cash flow
13.55   Completed financial cash flow
13.56   Completed project summary sheet
13.57   Completed project sensitivity analysis

1   Introduction

Scope and purpose of the book

This book sets out to provide development-modelling solutions for the twenty-first century. In this it might appear to be little different from the other traditional titles but actually it has a different philosophy from the books that have preceded it.

The approach reflects the fact that anyone setting out to do a development appraisal will not use the traditional approaches using a simplified manual calculation with valuation tables but will use a spreadsheet or proprietary software system. All the existing development appraisal books have their origins in the 1970s and 1980s and are thus outdated. This book takes a very fresh approach, illustrating development modelling from theory to actual current practice.

Using spreadsheets and proprietary models has greatly extended the appraiser's ability to carry out complex development appraisals yet the models that are used have often been produced on an ad hoc basis by trial and error, which in turn leads to the potential for error. This text aims to fill in the gaps between theory and practice by illustrating the solutions to many typical appraisal problems faced in practice using the tools that the industry itself would use.

Three solutions are considered; Excel, Estate Master DF and Argus Developer, although, for reasons that will be outlined in the text, concentration is made on the latter two models.

The structure of this book

The book is divided into two parts.

The first part looks at the theory of development appraisal and the construction of feasibility studies. It looks briefly at the nature of development, the parties involved and the role of development appraisal in the development process. Then the process of building the appraisal commences, starting with a brief look at the goals of developers and how this affects the appraisal. A key area is then considered: an examination of the factors that constrain a development, essentially those factors that define the envelope of the scheme – this is not just confined to the physical envelope but the use type and potential users and occupiers as well. This is important: development is a creative process; the developer has to use their imagination to visualise what might be built on a particular site or in a locality. Often this concept is already defined by planning and legal constraints but frequently it is not and it is often left to whoever is doing the initial appraisal to determine the approximate envelopes.

Once this is done a broad idea of the overall development scheme will have been established. The next step looks at the gathering of the data that will construct and inform the appraisal itself. A wide variety of information is required and whilst a fully comprehensive list is impossible in a single text given the varied types of property and development that takes place, the reader will gain a good feel for the type of data gathering required and, perhaps more importantly, the reasons for its collection.

The final section of part one looks at the mechanics of the construction of development appraisals themselves. This starts with an examination of the simplest and most traditional of the approaches, the residual approach, and then moves onto cash flows. It concludes with an examination of an essential component of development appraisal, sensitivity analysis.

Part two of the book moves from the abstract and theory to the practicalities of development appraisal in the twenty-first century. Virtually all appraisals are produced using computers, either using spreadsheets such as Excel or using one of the proprietary systems such as Argus Developer and Estate Master DF. This part is, therefore, split into three, the first covering Excel, the second devoted to appraisals using Argus Developer and the final part constructing appraisals using Estate Master DF. No dedicated text exists for either of the last popular systems and it is felt that both students and practitioners will gain an insight into both systems by the examples illustrated here.

I would like to acknowledge the kind permission of Argus and Estate Master for the use of the images used in Chapters 12 and 13 respectively, and, indeed, for allowing me to write this book with total editorial freedom.

Part One

Principles of development modelling

2 The background to development appraisal

Definition of property development

Before we can start on a book that concentrates on development financial feasibility studies we need to establish some definitions.

The first is to try to define what property development is. You might think that is simple; property development is about building things surely? Well, whilst that is often true this does not encompass all development forms; some development involves no new construction but the adaptation and/or alteration of an existing structure. Whatever, this is clearly development. Similarly, many entry level projects in the residential markets involve little more than redecoration.

'Development' is defined by statute in the Town and Country Planning Act 1990 s55(1), as 'the carrying out of building, engineering, mining or other operations in, on, over or under land, or the making of any material change in the use of any buildings or other land'.

That is quite a wide definition, but does little to actually define the activities involved. Wikipedia is quite helpful in this regard. Its definition is:

> Real estate development, or property development, is a multifaceted business, encompassing activities that range from the renovation and re-lease of existing buildings to the purchase of raw land and the sale of improved land or parcels to others. Developers are the coordinators of the activities, converting ideas on paper into real property.
> (http://en.wikipedia.org/wiki/Real_estate_development, accessed February 2013)

In 2002 and in the second edition in 2008, in *Contemporary Property Development*, the definition I used was:

> Property development is taken in general to be the process that involves the transformation of property from one state to another.
> (T.M. Havard (2008), *Contemporary Property Development* (2nd edn), London: RIBA Bookshops, p.1)

The point I was trying to make then and which is still as valid now is that property development is a broad subject. There are many different types of property development projects. As a result every development appraisal is also individual.

The property markets and development

Laymen and the media often refer to the 'property market', treating it as a single entity. This is of course far from the case; there are in fact numerous property markets, in fact it can almost be infinitely sub-divided. The residential market is distinct from the commercial. Residential property can split into its sub-types; detached houses, semi-detached, terraces, flats, houses in multiple occupation (HMO) etc. Each type can be split into user/occupier type – retirement, single-family, starter homes, urban professionals, second homes, student lets etc. A further sub-division is possible by way of location – country level, region, town/city, suburb, street etc., all of which can influence value, saleability and market performance generally. When you add the fact that several interests can exist in the same property – freeholder, head lessee, tenant, sub-tenant, licensee etc. – then some of the complexity can be understood.

Each of these markets can and will behave differently. Each will have their own market dynamics, with different supply and demand balances and, potentially, their own development markets. It is well beyond the scope of this book to provide a full analysis of the property markets; however, an appraiser/developer must have a good feel for this area if they are to be able to efficiently produce accurate development feasibility studies.

Property developers

Similar to the above section, it is largely beyond the scope of a book that concentrates on development feasibility studies to cover the development sector in depth; however some coverage will be given here.

Residential development in the UK is dominated at the volume end by the major house builders and construction companies. According to www.theconstructionindex.co.uk, at the end of 2012, the twenty leading companies in the UK were as shown in Figure 2.1.

Beneath this list, however, are numbers of small building companies who undertake development and myriad small developers undertaking single refurbishment projects (but which are also classified as development). Whilst volume house building is a major industry requiring major investment, the small-scale residential refurbishment represents the entry level into the industry.

Also in house building there are the social housing developers, currently dominated by the housing associations. These not-for-profit organisations are significant players in the housing market having bought up former local authority stock but also commission considerable numbers of new build properties (see Figure 2.2).

Rank by turnover	Rank by profit	Company	Latest turnover (£m)	Previous turnover (£m)	Latest pre-tax profit (£m)	Previous pre-tax profit (£m)	Latest margin (%)	Previous margin (%)
1	18	Barratt	2,035.4	2,035.2	-11.5	-162.9	-0.6	-8.0
2	3	Taylor Wimpey	1,808.0	2,603.3	78.6	-71.3	3.0	-2.7
3	2	Persimmon	1,535.0	1,569.5	147.2	153.9	9.4	9.8
4	1	Berkeley	1,041.1	742.6	214.8	136.2	28.9	18.3
5	4	Bellway	886.1	768.3	67.2	44.4	8.7	5.8
6	6	Redrow	452.7	396.9	25.3	0.5	6.4	0.1
7	5	Galliford Try[2]	388.5	316.0	31.6	17.6	10.0	5.6
8	7	Bovis Homes	364.8	298.6	23.3	14.0	7.8	4.7
9	19	Crest Nicholson	319.1	284.4	-27.0	-27.4	-9.5	-9.6
10	8	Bloor[1]	300.9	312.0	12.2	16.0	3.9	5.1
11	10	Countryside	267.5	228.1	9.9	-0.4	4.3	-0.2
12	16	McCarthy & Stone	230.6	203.3	1.3	-8.7	0.6	-4.3
13	17	Stewart Milne	229.1	250.8	0.4	0.1	0.2	0.0
14	20	Miller	220.9	261.3	-33.2	-98.7	-12.7	-37.8
15	15	Cala	214.2	114.9	2.0	-27.1	1.7	-23.6
16	12	Hill Partnerships	178.9	133.5	6.5	6.0	4.9	4.5
17	14	Kier	153.0	158.0	4.2	2.8	2.7	1.8
18	13	Morris	136.0	126.1	5.6	1.3	4.4	1.0
19	11	Fairview	129.4	97.1	6.6	-5.3	6.8	-5.5
20	9	Keepmoat Homes	126.0	70.8	11.4	3.8	16.1	5.4
		Totals	11,017.2	10,970.7	576.4	-5.2	4.9	-1.5

All companies ranked according to their most recently filed accounts.

Notes:

1. *Bloor Holdings only publishes an operating profit (before exceptional items) for its house-building business.*
2. *Galliford Try profit figure is for operating profit.*

Figure 2.1 The 20 leading house builders in the UK. This table is copyright The Construction Index and is reproduced with permission

Development of commercial property has similar characteristics, with the top end of the market being dominated by large, often publically listed companies but with smaller builders and developers carrying out small refurbishment projects on a one-off basis.

It is hard to be definitive about the larger players in the market.

Some of the biggest private property companies such as Grosvenor, Portman and Cadogan Estates have their roots in landed estates but have diversified into wider property development and investment companies.

One of Britain's largest property companies, British Land, was founded in 1856 as an offshoot of the National Freehold Land Society. The main object of the National Freehold was to facilitate the acquisition of small plots of land by the people in order to qualify them to vote. With electoral reform, the company was reformed and began to operate as a normal business in the latter part of the nineteenth century. In 2011 it held more than £4.2bn of assets.

Land Securities is the largest commercial development and investment company in the UK. It was founded only in 1944 but grew rapidly and is now a Real Estate Investment Trust (REIT) with a FTSE100 listing.

4-year pipeline rating 2011	Rating 2010	ASSOCIATION	How many homes do you own/manage?	Homes expected to build between 1/4/11 and 31/3/15?
1	22	Sanctuary Group	79,011	8,000
2	2	Affinity Sutton	56,000	6,100
3	1	L&Q Housing	61,994	5,900
4	16	Sovereign	32,741	4,740
5	21	Bromford Group	26,995	4,500
6	8	A2 Dominion	33,815	4,400
7	6=	Notting Hill Housing Group	27,000	4,325
8	13=	Orbit Homes	360,000	4,200
9	19=	Swan Housing	10,826	4,141
10	19	The Guinness Partnership	60,500	4,000
11	10	Family Mosaic	23,000	3,844
12	11=	Hyde Group	45,000	3,700
13	9	Home Group	51,000	3,500
14	39	Network Housing	16,815	3,301
15	29	Waterloo Housing	17,500	3,150
16	18	Catalyst Housing Group	20,905	3,000
17	31	Housing 21	18,223	3,000
18	48	Midland Heart	32,000	3,000
19	15	BPHA	16,000	2,629
20	34=	Devon & Cornwall Housing	19,000	2,573
21	11	Metropolitan Housing Partnership	37,877	2,500
22	13=	One Housing Group	13,991	2,500
23	5	Southern Housing Group	26,000	2,350
24		Together Housing (Trans Pennine)	35,354	2,026
25	6	Places for People	61,926	2,000
26	3	Circle	63,000	1,904
27	33	Great Places	15,917	1,900
28	26	Thames Valley Housing	14,663	1,900
29	16=	Moat	20,740	1,615
30	40	Sentinel HA	7,794	1,600
31	46=	Aster Group	17,340	1,568
32		Longhurst Group	17,232	1,524
33		WM Housing Group	24,000	1,445
34		Gentoo Group	29,763	1,444
35	45	Somer Housing Group	11,961	1,440
36	49	Green Square	9,903	1,339
37		Viridian Housing	16,500	1,265
38		Arcadia Housing	11,000	1,200
39	50	Yorkshire Housing	15,378	1,200
40	27	Spectrum Housing	17,261	1,186
41	30	East Thames Homes	12,991	1,164
42	28	Derwent Living	11,000	1,139
43		Cross Keys Homes	9,838	1,080
44		LHA-ASRA Group	12,000	1,050
45		Accord Group	10,824	1,017
46		Fabrick Housing Group	15,031	1,000
47	34=	Paradigm	11,700	1,000
48	24=	Radian	18,092	988
49	40=	Town & Country HG	8,357	900
50		Vela Group	17,349	847

Figure 2.2 Housing associations in the UK in 2011

Source: insidehousing.co.uk

Many commercial companies such as Boots have, at times, had property investment and development arms. In addition, some life assurance companies such as Prudential (now PRUPIM) and Standard Life also undertake direct development projects, generally funded by equity funds (see the section on financing in Chapter 6).

Beneath these giants are smaller, more specialised development companies, some private and some listed, some funded privately, some relying on debt financing.

The development sector is large and varied, but all generally appraise development projects using the same techniques.

Changes in the development environment

Property development as an activity has seen considerable changes over the last 25–30 years, and these changes have had an impact on appraisal practice and techniques. Although this book is going to primarily concentrate on the techniques and tools available to the appraiser, I feel it is important to understand the context in which they are used.

Firstly, development has become broader both in its scope and in the type of people who are involved with it. With the encouragement of lifestyle programmes and the easier availability of finance (up until the credit crunch of 2008 at least), more and more people have been carrying out forms of development, the normal model being the purchase of a run-down house or flat, spending a few weeks remodelling, refitting and redecorating and then either letting and holding or selling on. This has brought a number of people of very differing backgrounds into the development and investment industry, which was, until then, dominated by builders and professionals.

Many of these 'new' developers do not carry out detailed appraisals even though, often, they perhaps should. They were cushioned in many respects by the almost continuous rise in house prices in the UK that was only arrested by the credit crunch and severe economic downturn of 2008. Until this time it was actually quite hard *not* to make money out of turning property in this way. The new, more difficult market may require these developers to take more care in assessing development projects. It seems increasingly likely that they will turn to more professional approaches, which will include the use of proprietary software.

The second big change in development is that residential schemes have moved out of the almost exclusive fiefdom of the house builder and into the mainstream.

It may seem odd but until quite recently in the UK the vast majority of private house building was carried out by specialist volume producers who did little else in the property market. Surveyors and other property advisors had little involvement in this market other than the acquisition of sites and the disposal of the end product. These house builders usually built up land banks of sites with potential for volume house building, worked at getting planning consent for them and then brought them on stream several years down the line to fit in with an overall production schedule. Individual houses were very keenly costed and produced at a budget. It was perhaps the closest thing the UK property market has seen to an industrial type production process. The housing development market was very specialised and almost a closed world to those outside it.

What changed all this seems to have been the urban regeneration movement and the increase in fashion for urban living.

Having spent most of the twentieth century seeing wealth and populations migrating to the suburbs and fringes of the urban areas, the last decade of the twentieth and the early ones of the twenty-first have seen a sharp reversal of this trend. Partly this is down to investment in and thus a greatly improved urban environment. Other factors include fashion, the development of late night/early morning entertainment in city centres, increased wealth and an increasing interest in buy-to-let investments opportunities from private investors. In addition, the rise in energy and transportation costs has made living closer to work more attractive. This latter trend seems likely to strengthen in the coming decades; the days of cheap energy and transportation seem to have gone.

Whatever, this trend means that development models have had to be able to deal with the special requirements of residential development. These include staggered starts and completions, deposits, buying off plan, complex disposal patterns and the like. These components are uncommon in commercial projects and this has had quite major implication for practitioners in the field. Most development appraisal books (and sections in books dealing with development appraisal) from earlier times have concentrated on commercial development for this was where the bulk of the work was for professionals. Now the picture is very different and advisors must be able to model all types of schemes.

Development and development appraisal has also shifted from being a local concern to an international activity. Of course development has always taken place in every country in the world but generally this used to be a primarily local activity carried out by local contractors and developers. Now increasingly, with cross border investment and the rise of international advisors and professionals, many developers, surveyors and development specialists are working across borders in many different countries. The methods and models used must therefore have applicability in many countries and not just one.

Development projects have also seemed to get much more complex. I say 'have seemed to' because this may be an illusion; developments have always involved complexity, it is just that our ability to model them has become much greater in the last 20–30 years with the advent of computers, spreadsheets and development appraisal software.

In the 1970s and 1980s, which is when the majority of development appraisal texts were written, the vast majority of appraisals would have been undertaken using valuation tables, perhaps aided by pocket calculators. To carry out the appraisals in a reasonable timescale the calculations had to be simplified (and also therefore made less accurate). This is why the residual appraisal technique (described more fully in Chapter 8) came to be so entrenched in the profession; it is nothing more than a simplified cash flow used to ease the burden of calculation. Once established, it has been hard to shift. Today, however, we are in an era where the tools that are available to us have become infinitely more powerful and our understanding of the modelling process has been greatly increased.

I think is important to appreciate all of these trends and the impact it has had on the appraisal calculation process. The market has become wider, more complex, more international and less parochial. The days of doing manual calculations using valuation tables have gone. The trends have pushed us towards cash flow models that are constructed on computers, which is the primary focus of this book.

3 The theory of development financial feasibility studies

Purposes of development appraisal and its role in the development process

Financial feasibility studies are also known as development appraisals and involve the gathering together of a range of diverse information on the costs and values of a project. This information is incorporated into a framework and used to determine the answer to some key questions. Development appraisal is one of the key aspects of assessing the viability of a development project. It is however used throughout the development process to fulfil a number of key tasks.

Primarily, appraisals are used to determine the price to be bid for a piece of land. A piece of development land has no intrinsic or set value; it only has a value derived from the use it can be put to. This is determined by the market and restrained by the planning and/or building restrictions on the site. Every scheme proposed for a site will generate different values. If a piece of development land has been fully exposed to the market then a range of appraisals (valuations) based on different schemes will be made on the site by different prospective developers. The landowner will generally sell to the developer who submits the highest viable bid. In this it can be seen that the financial appraisal is a key component in determining the highest bid a potential developer can make whilst still meeting their target return for the project.

A second major use of development appraisal in the development process is to determine the profit or loss the scheme will make. This is vitally important; it shows the developer whether the scheme is viable or not.

A variation of the profitability calculation can also be used to explore the impact of different variations in the project. These variations can be down to design changes, change of use type or use mix that the site can support, or changes to the timings of elements of the project. They can also be used to determine the point of peak profitability of the scheme.

The financial feasibility study will also be used by commercial lenders to determine whether money can be lent on the scheme. Commercial lenders will look at the financial appraisal very carefully before advancing any funds. In essence they are looking at two things:

- Whether the assumptions made by the developer in the appraisal are sound. They will carefully examine all of the components of the completed scheme. They will look at the rental values and the yields that have been projected by the developer in the appraisal for realism. They will also view the selling or leasing program to determine whether this can be achieved based upon the current market conditions. They will also examine the construction costs and all the other elements of the project. The appraisal lays these factors completely open to scrutiny. The developer will use the development appraisal to prove whether the assumptions in the development project are soundly based.
- If the financiers are satisfied with these factors they will then look closely at the profit margin projected in the appraisal. The financiers want to be satisfied that the developer will achieve a sufficient profit margin. This may seem surprising, but they are primarily concerned about the developer's financial stability. The profit margin reflects the risk margin on development. Basically, the larger the profit margin is, the less risk the lender will be taking in advancing funds on the scheme.

There are no set margins as to what lenders will look for in terms of returns. The normally accepted rules of thumb are for a 20 per cent profit margin on costs or speculative commercial schemes, 10 to 20 per cent returns on cost for commercial projects with pre-lettings, and 10 to 15 per cent on residential projects.

These are the principal uses of development feasibility. There are others as well.

As noted above, developers will explore the effects of altering, reworking and changing the timings in the scheme. Projects often require rethinking during the project lifetime. This may involve changing the mix of property that will be developed to suit market requirements. The appraisal will be used to see what the effect of these changes is on developer profitability.

The process of development analysis involves many disciplines of knowledge. The feasibility report should communicate the facts, assumptions, figures and recommendations gained during the analysis process. The person preparing this report need not have direct knowledge of all details required but they should at least understand the impact of each critical factor and how and where this information can be accurately determined. This process of information management requires investigative skills, mathematical application of data and intuitive thought.

Development appraisal is carried out at different stages and for different types of projects and, although the basic mechanics of appraisal stay the same, the character of the calculations are rather different; think of a sparrow and an eagle. They are both birds with the same basic features – wings, feathers, a beak – but the details and make-up are very different. This analogy applies to development appraisals; some of the non-standard forms of appraisal may seem different but the underlying fundamentals are the same.

Most development appraisals are done on 'for profit' schemes of a relatively short-term nature, and this is indeed what this book focusses on. The use of 'short-term' might seem surprising since many development projects take many years to complete their full cycle; however they are short term when compared with 'for profit' infrastructure type projects such as toll bridges and toll roads. Development appraisals can be done on these projects as well but the character has to be different taking into account the long time-scales, payback periods and on-going asset maintenance. Similarly some projects are 'not for profit' – things like public infrastructure schemes. Even these use a form of development/investment appraisal, monetising otherwise non-monetary items in a benefit-cost analysis. Although the mechanics are essentially the same we will be concentrating on the mainstream, shorter projects for the appraisals in this book.

Even within this area there are recognisable differences depending on the point at which an appraisal is carried out. It is a point that is perhaps arguable, but there are three distinct stages in a development when appraisals are carried out:

- Initial viability studies
- Detailed/final viability studies
- Project management monitoring studies.

These are carried out at different stages in the project and have differing purposes.

An *initial viability study*:

- is carried out to test whether a development is viable.
- tests whether the proposed development is the highest and best use for the site (i.e. that the value on completion is envisaged as being higher than the combined cost of purchasing the site for its existing use value, the cost of construction (including all fees and interest) and providing an acceptable profit margin/return to the developer).
- is unlikely to contain any scheme design at this stage, just the broad outline or concept of what scheme might be developed. The appraiser must therefore make broad, general assumptions.

The appraisal is unlikely to produce more than a general indication that the scheme is viable. However, the more details that can be established, the more accurate this appraisal is likely to be.

A broad indication that a scheme is viable may lead to the developer commissioning outline designs and gathering other data; however these slightly later appraisals would still be classified as initial appraisals.

The *detailed/final viability* study often overlaps with these later initial feasibility studies; indeed the division between the two is both blurred and arbitrary. However most practitioners would agree that they exist as separate entities.

- The initial viability study gives a basic yes no answer and also possibly addresses the question of when. If there is a yes at the preliminary stage of the development project then moves to the next stage. This is significant for the developer because it will involve a higher level of costs. These costs may include:

  - The acquisition of land or acquisition of an option to buy the land at a future date.
  - The scheme design. This will either require the employment of an architect or the assembly of an integrated construction team (design/build etc.). Design and build has advantages (speed, reduce cost, lower client management input etc.) but there are major disadvantages for certain sectors, particularly the commercial investment market. These disadvantages include less control of the detailed design and subsequently control over the completed product. This usually pushes this sector towards the traditional route.

- This stage may also see the employment of other consultants. This will add detail to the scheme and make many of the estimated components in the appraisal more soundly based but will add costs:

  - QS/cost consultant/construction economist. These will analyse, monitor and advise on construction costs at the design proceeds received.
  - A planning consultant or consultants to advise on negotiations with the planning authorities.
  - A market analysis/market consultant to monitor market requirements.
  - Some schemes may also require historical/ontological/conservation advice.
  - Most developers will also take on legal advisers.

The appraisals done at this time should be more complex and sophisticated than the initial appraisals. They should also be more accurate. This is all due to the increase in information about the scheme. They will be used to continually monitor the effects of the solidifying design and cost elements and on the profitability/viability of the scheme. If these are the internal aspects of the project, the appraisal also allow the monitoring of the effects of external aspects of the project, i.e. the ones that are out of the control of the developer. These things include how the national and local economy is doing, how prices, values and rents are moving and also what the competitors are doing. Market opportunities will be spotted by more than one potential developer, and new schemes may come to the market whilst the developer is preparing their own project.

The key issue is that there is a final yes/no point that all developers are aware of. The decision may be to proceed immediately or delay the start. Whatever the decision, a point occurs just before the major expenditure takes place. Once the building contract is in place it is virtually impossible to stop the development being built out. Even where a site has been acquired, a halt at this point is not as financially damaging as building out a scheme that cannot be let or sold. Before the development starts, the existing, presumably low value use (as an old office building, storage or car park for example) can continue to offset some of the cost. If necessary, it may be possible to sell the land to another developer and recoup most of all of the initial outlay.

For these reasons, the period immediately before letting the building contract is the important break point in the development. It is the point where the final development appraisal will take place that will inform the final key decision for the developer as to whether to go ahead or not.

In days gone by, the initial appraisal might have been done using a traditional residual approach whilst the later one would have been worth employing the greater investment in time and effort of a spreadsheet/cash flow approach. Those appraisers still using self-created spreadsheet models will probably still use this division; however users of the proprietary systems such as Argus Developer and Estate Master DF will use the same models, just adding more detail as the project proceeds.

Project management monitoring studies are the third major use of feasibility studies and essentially involve using the final appraisal as the plan against which the real data (expenditure and receipts) is recorded, essentially reconstructing the appraisal in 'real-time'. This is the best way of ensuring that the project is remaining on track and also helps the developer most efficiently plan their financing requirements for the project.

4   The basic development appraisal equation

Before we turn to the models in detail, it is important to look at the basics of development appraisal.

The basic equations for a development appraisal are simple. To calculate development land value (or rather a land bid) the equation is:

Value of the buildings on completion

Less:

The development costs (Construction, all fees, all ancillary costs,
and all the costs of finance etc.)

Less: An allowance for developer's profit

Equals

Land Value (Maximum sum available to buy land)

The alternative equation is used when developers know their likely input costs for land and construction and want to discover whether the scheme is viable, i.e. whether it produces sufficient profit for them to proceed:

Value of the buildings on completion

Less:

The development costs (Construction, all fees, all ancillary costs,
and all the costs of finance etc.)

Less: Land Cost (including fees)

Equals

Development Profitability

It will be noted that the appraisal is always looking to solve for the unknown element in the equation, either land value or profitability. All of the above elements will have to be established or estimated.

The complexity in appraisal comes from ensuring that everything has been accounted for, calculating their correct value and allowing for when these items will take place, which in itself will impact on their value. This process is further complicated by everything occurring in the future with the developer/appraiser often dealing with a scheme that only exists in outline. A development appraisal is always no more than a forecast of a series of future, uncertain events.

When, later in this section, the process of actually putting the appraisal together is examined in more detail, it is easiest to look at this process by looking at the traditional way of doing development appraisals, the residual approach, because this approach clearly follows this three section model. Cash flow (and the proprietary systems) essentially do the same; however their more complex structure sometimes disguises this, hence we will look at these models after the residual approach.

It should also be noted that the land residual sum will have to be adjusted for the effect of time, essentially discounting the sum back to the present in both cases. The process of constructing the appraisals and the principle of discounting are explained in Figures 4.1 to 4.4.

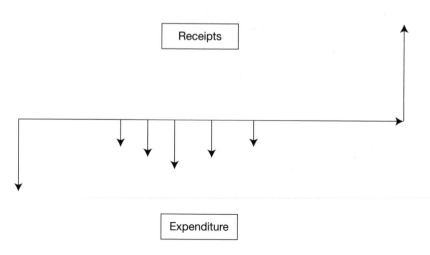

Figure 4.1 Development appraisal involves several steps. The first is to assess what the costs will be to develop the site (this will include the purchase of the land, the site preparation, the building costs and professional fees etc.) and the receipts from the sale or letting of the buildings produced. The second thing is to determine how long the process will take and when the expenditure and receipts will take place. This process is presented in this diagram, which represents a simple development with time running from left to right. This development starts with the major expenditure of purchasing the land. Monies are then spent on developing the site and then finally these buildings are sold. Although this is a simple project it is representative of the tone of cash flows in many projects.

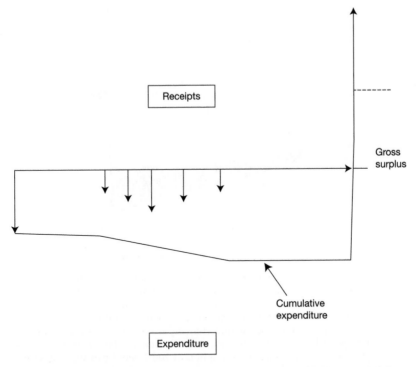

Figure 4.2 In the appraisal itself, the next step is to total up all the costs and the
receipts. This accumulative process is typical of most development
appraisal models.

Another thing to note even at this point is that the relationship between the
residual outcome is highly geared and very sensitive to the assumed inputs. This
is a crucial element to understand, and can be examined with some simple examples.

Let us use the following project, whose initial appraisal was as follows:

Value On Completion	£10,000,000
Development Costs (inc. interest)	(£6,000,000)
Land Cost (inc. holding costs and interest)	(£2,300,000)
Development Profit	£1,700,000

This is a profit of 20.48 per cent on costs.

If, between doing the appraisal and the development being completed, values
fall 5 per cent, the following happens to the figures:

Value On Completion	£9,500,000
Development Costs (inc. interest)	(£6,000,000)
Land Cost (inc. holding costs and interest)	(£2,300,000)
Development Profit	£1,200,000

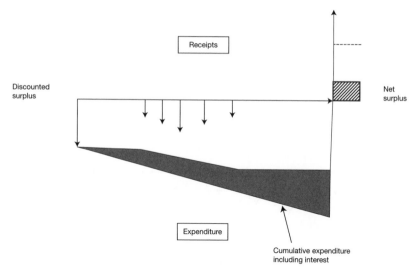

Figure 4.3 The next step in the appraisal process is to account for the effect of time
on the project viability. In most cases this is done by calculating the
interest charge that will be accrued on the development expenditure.
In practice this interest may have different values on different elements
of the project (land, construction etc.) and from different sources (equity,
principle debt, mezzanine finance etc.). To simplify the calculation, many
appraisals are done with an assumption of 100% debt financing and with
an equal cost (see the comments on opportunity cost in text). In addition
the traditional residual approach to appraisal makes a gross simplification
in interest cost calculation (see later section). One of the key advantages
that both the cash flow and the proprietary models have over the
traditional residual is that they calculate interest more accurately (subject
to the underlying cash flow being forecast accurately). The surplus at the
end of the project is now net of interest.

Although this is still a 14.45 per cent profit on cost, it can be seen that the
5 per cent drop in values, something which can easily happen over the period
from inception to completion of a development, has been magnified into a
30 per cent drop in profitability. This is worrying on its own but often
a deterioration in values is accompanied by an increase in length of time to let
or sell the scheme. This increase in time increases costs, essentially due to a
rise in interest charges. If we look at a 10 per cent drop in values combined
with a 10 per cent increase in costs, this has the following effect on profitability:

Value On Completion	£9,000,000
Development Costs (inc. interest)	(£6,600,000)
Land Cost (inc. holding costs and interest)	(£2,530,000)
Development Profit (Loss)	(£130,000)

The scheme now makes a substantial loss.

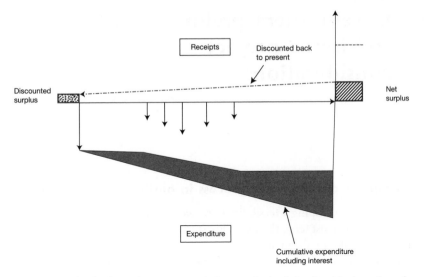

Figure 4.4 The final step in most appraisals, true for both land residuals and profit calculations, is to discount the future surplus back to the present. In both cases the surplus is a future value, only available in the future. Given the long nature of development projects this can be several months or even several years in the future and as £1 paid at any time in the future will always be worth less than £1 today, discounting is required. Normally this discounting is done using the finance rate of interest.

The lesson of this is twofold: Firstly, the highly geared/sensitive nature of development means that only small market movements can turn a project not viable. This is one of the reasons why development is a high-risk activity and also why appraisals and research throughout the period running up to the development start are essential. The second lesson is how accurate the appraisal needs to be; the outcome above can also be achieved with no deterioration in the market but by the appraiser/developer being slightly optimistic about sale or leasing prices and underestimating the cost or timing elements. It underscores the need for the inputs to the appraisal to be as accurate and carefully considered as possible, which is the purpose of the next section of this book.

This also seems a good point to introduce a criticism of proprietary systems. This book is a strong advocate of such systems because of their reliability and their ability to greatly extend the modelling ability of developers and appraisers but one, quite valid point, against them is that the ease of use of the systems makes it rather easy to produce an appraisal without proper consideration of the validity of the inputs. This is particularly important when the appraisal is being constructed by someone who is relatively inexperienced. The above examples should be a cautionary warning. If the classic old-adage of property is 'location, location, location' then for development appraisal it should be 'validate, validate, validate'.

5 Development preliminaries and initial project considerations

The property development process in outline

The process of developing a feasibility starts well before the project itself starts, indeed, as we have just seen, the feasibility study essentially determines whether the project should go ahead at all.

Establishing project objectives

The first step is for the developer or appraiser to establish the project's objectives.

These are very important things for both the developer and appraiser to establish, as they will set the key benchmarks that will have to be achieved for the development to go ahead.

Different types of development objectives

Just as it is hard to define all of the types of valuation that exist it is equally as hard to produce a definitive list of what objectives a developer has.

Feasibility studies are strongly associated with private sector developers. Financial feasibility studies however have wider applications. Some projects are not undertaken for purely financial reasons but sometimes for other, perhaps social factors for example, that they produce. These will be considered in more detail in the subsection below, however they can include the provision of employment and the construction of infrastructure, enabling other developments to take place. Essentially the same techniques can be used in the appraisal of these projects but they do tend to be carried out using wider cost/benefit analysis. These look at the wider costs and benefits to society as a whole. Strictly, they are outside the scope of this book; however this process is normally done by converting non-monetary items into monetary terms with the figures applied in a loose development appraisal format.

In terms of other, more mainstream, market projects, it is important to the developers to have a reasonably clear idea of what they're setting out to do before they carry out a financial appraisal. Most developers will be aiming to maximise their profits; however there are other financial goals that a developer may seek. These include the creation of a long-term income stream and the carrying out of, initially, non-economic development which will enable later, economic, phases to be carried out.

In addition to the financial aims, the developer will need to define where the development is positioned in the marketplace. For example, the developer will need to decide whether the development is aimed at the luxury or entry-level in the market. Also, whether the developer is producing properties to be retained as an investment or if the intention is to sell on to either other investors or to the owner-occupier market. Also fundamental is the use type which the developer is aiming at. Examples include retail, office, industrial and so on. Each of these will also require determination of the segment of the market at which the project is aimed. As an example, the office market ranges from simple low-cost, low rent suites and rooms to the complex, state-of-the-art, luxury headquarters type buildings. The range of costs, rents and values are very wide, and will obviously influence the feasibility study considerably.

All these factors will influence what site is selected and the final form of the development, but also will define the benchmark targets the developer will set.

The developer must, therefore, have a clear idea of the type or types of development they are aiming to produce and the segment of the market they are aiming to serve prior to commencing the detail planning of the project. Note, however, that the market analysis, covered in the text below, can inform the development of what this focus should be as part of the financial feasibility study.

Social objectives

Social objectives are normally the preserve and interests of public bodies, local and/or national government and other bodies to which the social objectives have been devolved to. Examples of the latter include housing association and quangos (quasi autonomous non-government organisations), local healthcare trusts and charities. All of these bodies can and do carry out projects and all will need to carry out development appraisals/financial feasibility studies.

Most will not have the objective of maximising profit, but most will be accountable and want to maximise the value of money whilst meeting the social objectives that they have set themselves.

These objectives may include the provision of affordable housing for key low-paid workers and providing care and health facilities to meet the needs of the community. The appraisals carried out by these organisations use the same techniques as those used by more general developers but do tend to be rather specialised.

It should be noted that there is an increasing trend for Public-Private Partnerships (PPP) or Private Finance Initiative (PFI) projects, where the private sector provides buildings (often serviced/staffed) for the public sector. These include the provision of hospitals, prison, schools and office buildings.

The private sector provider certainly will have the profit motive in mind whilst doing the appraisal but, again, the financial feasibility studies tend to be special-ised, extending further into the future and including non-construction type expenditure.

Very occasionally private sector developers will also have social objectives in their current appraisal or, more likely, will have to incorporate social goals such as the provision of affordable housing into their feasibility studies at the behest of a landowner or planning authority.

Financial objectives

Financial objectives often overlap with the overall development objectives in a private sector development. Social developments by non-profit organisations can, however, also have financial objectives such as achieving cost minimisa-tion and/or value for money. Both of these can be assessed using a financial feasibility study.

In terms of market-driven developers, the normal objective is for profit. This is either measured as a straight profit – usually the net present value or the returns on cost. Alternatively there are various yield measures. These include income returns, either on cost or on end value, or on some form of internal rate of return (IRR).

As we have already alluded to, it is normal for the developers to work to benchmark targets. Common benchmarks include certain levels of return on cost. For example a typical return on cost and commercial development project is around 20 per cent profit on cost. The residential benchmark has traditionally been lower, often in the range of 12 to 15 per cent, as these projects are perceived as being of a lower risk (the end user market for residential property is generally taken to be wider than for any commercial property – everyone needs somewhere to live but not everyone needs business premises). These benchmarks, however, do vary over time. I will return to this again and again as we work through financial feasibility studies, but essentially what is an acceptable return should be project specific and dependent upon the degree of risk involved. A low-risk project of any type need not have a high margin whilst even normally mainstream, popular types of developments, such as flats/apartments, may require high margins if there is something about the project – the state of the market, proximity to a proposed road scheme, or a high-speed train route, for example – that creates unusual uncertainty should see the profit (risk) margin set higher.

6 Establishing development constraints – the development envelope

Once the project objectives have been defined then the developer/appraiser should consider the potential constraints on development.

Some constraints will put a complete bar on the development taking place at all but most will establish the limits of the development – the physical limits, legal limits and market limits. This is very important in defining the development form.

This function of the appraiser is sometimes overlooked. In the built environment, it is generally assumed that it is the architect who defines the built form. This is true in detail but, prior to this, it is often the developer/appraiser who has to consider the basic question of what can possibly be developed and where. The development feasibility study has a key role in this process.

In some cases, the potential development envelope is vast, the boundaries ill defined. A good example of this was Dubai and Abu Dhabi in the 2005–2009 period where virtually anything could be built. Both locations had vast tracts of cheap, empty land with few planning constraints, ready access to finance and a huge potential market from buyers, occupiers and investors who had bought into the concept. The result was an explosion of development and the exercise of many architects' and developers' imaginations that resulted in projects such as the Burj Khalifa – the tallest building in the world at the time of writing – and the World and Palm Island schemes.

Contrast this environment with that which faces the developer contemplating a project in the centre of a historic British city such as Oxford, Bath or Edinburgh's New Town. The envelope for the development here is going to be highly defined and constrained; essentially the scope for variation will be limited and the developer will have to establish exactly what these constraints are before they can accurately assess a scheme.

Most development projects will fall between these extremes but it is important to recognise first the type of constraints that exist, that they go way beyond the simple restrictions that the planning system places on development and, also, that it is developers and appraisers that have to do the initial formulation of ideas and the potential for development that will only later be realised in architects' plans and in bricks-and-mortar.

This next section explores these constraints in more detail.

Planning

Development planning consent is clearly very important. The UK has a highly constrained site-by-site detailed land use planning system. Other countries have more liberal zoning systems that allow the site to be developed for particular type of use without defining the details. Whatever the system, all define the land use and some if not all of the detailed form of the development. This includes the size, form, building restrictions, often the materials used and sometimes the detailed specifications. These are of course all the elements that affect both cost and value.

To give an example of the impact of control, consider a site that has consent for a decentralised office park. In the UK the planning authorities will approve the final drawings when they give detailed planning consent but even at the outline consent stage the planning regime will define the floor space ratio. For example a 1 ha site may allow a 35 per cent build ratio with a three-storey maximum build height. This will essentially set the size and form of scheme that can be built in this location.

In established built-up areas, particularly in conservation areas in city centres, there may be requirements to retain elements of the existing building on the site. It is quite common in sensitive historic areas to require the retention of the facade. In some very sensitive locations, examples of which include parts of the new town of Edinburgh, historic towns such as Oxford and parts of Bath, sometimes whole rooms visible from the street level have to be retained in the development. This of course has an impact on cost – the need to use historic building materials, the inefficiencies of working whilst retaining elements of the building, etc. – and on value – restricting the ability to maximise the amount of lettable space that can be developed.

A whole book could be written on the planning system and its effect on the development process. It is a very important area of development; however a detailed examination is beyond the scope of this book, which is concentrating on the feasibility study itself. However the importance of this area cannot be overstressed, and the type of inquiries and investigations required from the developer/appraiser are considered in a later section.

Contractual

There are a great number of contractual issues that an appraiser has to investigate prior to doing a development appraisal. However, to do a development appraisal, you do not have to have any contract in place at all. An appraisal can be done at any time and on any piece of land, even one that is owned by other people. This is, actually, quite a common situation although these appraisals are often difficult to carry out as it is problematical to get accurate information without revealing the developer's interest.

There are a number of contracts however that do affect the appraisal and to which careful regard should be made. The most obvious is a contract to purchase the land. This defines the price to be paid, the timing (particularly the commencement of the scheme) and also the financial commitments as the contract may define the stage payments on land. These include the deposit, the final completion date, staged payments and any overage calculations that has to be included in the calculation. The land acquisition is a major cost and the timing of payments and their size has a huge impact on the development's profitability.

Other contracts which have an impact on the appraisal include the building contracts. If they are in place they will give structure to the construction expenditure, defining the timings and the amount to be expended. At the more detailed appraisal stage, therefore, it is important to look at the actual or the likely terms of the building contract itself.

If building work is being done at the time when building prices are fluctuating, then this must be factored into the appraisal. It must be determined whether the contract is likely to be a fixed-price contract or a variable one. A fixed-price building contract means that, if there are no variations in the scheme during the build period, the price agreed at the start of the contract will be the price is actually paid at the end of the development. This is highly desirable as it makes the costs very certain and predictable. If however a variable contract has been put into place, which is more common outside of the UK, then the appraiser would have to model the likely variation in building price as the contract proceeds.

Other contractual issues which an appraiser would have to take into account include contractual undertakings to do something as part of the development, examples of which include Section 106 agreements in England and Wales (see below) or the provision of social or low-cost housing as part of the scheme. All of this data has to be incorporated within the development and must be fully investigated by the appraiser beforehand. There may also be requirements to develop the site by a particular date, an imposed particular timescale from an outside agency which may restrict the developer's ability to maximise the profitability.

Other areas related to contracts to be developed and investigated as part of the appraisal include things like the title that has been bought. Does the vendor have the right to sell? If so what interests are being sold; is it the freehold or a leasehold interest that is being offered? If it is the leasehold, what are the details of constraints of the lease document?

For the implementation of the scheme itself the developer will have to determine where the boundaries are clear and well established. This is something that the legal team will investigate if the scheme is going ahead but this is also something the appraiser should have a mind to even at the initial appraisal stage. If there are potential problems and issues about land ownership and about boundaries, this may well delay the scheme and will certainly impact on the appraisal.

Similar aspects that will have an impact on cost and value are rights over the land. Again, at the initial appraisal stages these are very difficult to establish as these are often only revealed when the developer's full legal searches have been undertaken; however, in certain circumstances such as projects in city centres, they should be expected and factored into the appraisal. These rights over the land include easements – rights given to certain statutory authorities such as gas suppliers, water and electricity companies that gives them leave to run service pipes and cabling through the land and to have access to them to maintain and alter them. There are also rights of way and rights of support to consider as well as potential issues of rights of light. These areas are very difficult to establish but also quite difficult to extinguish, so the wise appraiser would always have the potential existence of these in mind. These issues can have a major impact on the scheme itself, both in terms of the cost of incorporating them, delays in extinguishing them and impact on the design of the scheme.

Other contracts which should be very carefully considered are contracts to let or sell the completed development (or parts of the development) prior to commencement. Although in some circumstances a commitment to sell or let it can be detrimental to a developer, particularly in a rising market, it also gives the benefit of higher financial security to the developer and any lenders on the scheme. Any contract or precontract signed which will sell or let a substantial part of the completed development will see any interest charges on loans being significantly lower than that achieved on a purely speculative scheme. This will reduce the development costs.

Political

Politics can have a strong influence on development and therefore the appraisal.

Development can be very controversial; developers have to walk a very narrow path to keep their neighbours onside. Some things might be developed perfectly legally but politics may require the scheme to be modified. The developer may have to adjust the scheme and the appraisal to produce a more politically acceptable product, particularly if they intend to do future work in the area which will require long-term goodwill of the community. Whilst it is hard to be specific about these factors again an appraiser must have one eye on the political situation and be prepared to adjust the figures projected accordingly.

Environmental

Environmental issues cover a wide variety of factors. For example, there are aspects that are external to the building which will impact on costs and values.

One external factor affecting the site is its presence in relation to floodplains or flood risk areas. Also there are other natural factors such as the building's exposure and vulnerability to wind, rain and weathering and also, in some areas, the building's exposure to the sun and its potential for solar heat gain issues.

Action taken to protect the building and its occupiers from these factors can impact on both cost and value. As an example of the latter, houses in flood risk areas in the UK have become increasingly uninsurable. This will impact on their market appeal and, therefore, on their values.

Other external environmental factors include man-made ones. They include things such as contaminated land and other issues of pollution. Any issues of contamination will have to be dealt with. Decontamination can be time consuming and expensive. The stigma of contamination can also impact on value.

Other environmental factors are to do with the building itself. There is increasing concentration on energy issues. There is a desire and indeed legislation to reduce the carbon footprint of buildings and their energy consumption generally. All new buildings have to meet stringent building regulations and regulations on emissions. These must be considered at the time of design; they will impact on cost, on the build period and also potentially on value.

Market timing

Timing is perhaps the most crucial thing in development. This applies to general and also the specific/special timings. Whatever, timing is a constraint that is often overlooked.

General market timing is about trying to ensure that the development hits its market on an upturn if possible. In most businesses, it is best to buy low and sell high and this is particularly true of property development land purchases. Because of the geared nature of the development appraisal (see the section above), land values can fluctuate sharply both upwards and downwards. Buying land close to the bottom of an economic cycle can see huge appreciations on an up-swing.

However, there are more specific examples of timing which could have an impact on cost (and therefore need to be accommodated in the schedule, requiring celebration of schedule or require working through inclement weather in order to meet deadlines) and also values. I can give three examples. The first affects retailing in Europe, USA, Australia and New Zealand. In all these markets the importance to retailers of trading in the Christmas trading period is critical. In fact, as retailers need to have time to fit out and to run their marketing campaigns in time for peak trading, this means that most retail schemes aim to be completed in September and October.

The second example is the hotel market prior to major sporting or other such events. These include the Olympics, football tournaments, City of Culture designation etc. These hotels need to be developed well before the event takes place. If they are late they are likely to be empty.

The third example also concerns hotels but those located in resort locations, either as a summer resort hotel, or a winter resort, i.e. ski lodges. These schemes will need to be completed prior to the peak letting time otherwise there will be a huge hit on value.

The nature of the development is that it is long-term but once elements of it start (land acquisition, placing the building contracts etc.), it is very hard to stop. Economic modelling has improved but no one has perfect foresight – if the author did he would not be writing about appraisal but INSTEAD would either be retired or STILL doing it! Everybody wants to hit the right timing for the development but sometimes this cannot occur. This underlines the importance of sensitivity analysis in the appraisal to ask the 'what if?' questions about timing. Sensitivity analysis will be looked at in detail later in the book.

Financing development schemes

Finance is one of the key components of development. It also often represents its biggest constraints. It is one of the key areas that developers and appraisers must investigate. Where the development finance is sourced, how the finance is structured and how much it costs is one of the key factors in appraisal.

Finance for development projects can come from a variety of sources.

Equity funds

Equity funds differ from debt sources of funding by the fact that they do not have to be repaid. Equity funds are normally internal but can be externally sourced, usually where the development has been presold to a property company, Real Estate Investment Trust (REIT) or financial institution and the receipts for the sale are drawn down to pay for the development costs.

The root source of these funds are:

* Commercial concerns using retained profits or shareholder funds to pay for building projects, usually for their own use but occasionally for sale or lease.
* Investment organisations such as pension funds and life assurance companies who invest the funds collected in premiums and in investment policies.
* Property companies who use shareholder funds or retained profits for reinvestment.
* REITs (Special tax status property investment trusts).

Debt finance

All funds that have to be repaid are classified as being debt. Development finance is scare, particularly equity sources, and many developers use debt finance in projects. There are also advantages to using debt finance. Firstly, in a rising market, increases in the scheme value will greatly increase the return on equity invested due to the geared nature of financial returns. Secondly, the money used in a scheme that is 100 per cent equity financed could instead have been spread over 3–4 projects giving more opportunity to gather profits (explanation: most lenders will only provide 60–75 per cent of the funding for a scheme, normally requiring the developer to provide the rest out of their own resources. Presuming lenders are willing to fund 75 per cent of a schemes costs, £1m could fund a single equity scheme and return £200,000 (20 per cent return on costs) or provide the equity seed funding to carry out four debt funded schemes, each also returning £200,000 each, giving a total return on equity of 80 per cent).[1]

Sources of debt funding include:

- *High street banks* – These are generally the primary sources of debt funding for development and the bell-weather of the health and viability of many development schemes. The banks have traditionally favoured lending to the sector because of the high margins on lending which they can obtain on the loans and the fact that when prices are rising, as they did in the residential sector almost continuously for more than a decade in the UK, the risks seemed negligible. The economic shock of the global financial crisis which started in 2007 showed that the risks were both real and high and development funds from this source have greatly reduced and the terms of lending tightened. All primary lending from this source will require the lending to be secured on the title to the site (generally). In good times it was possible to secure 100 per cent of the cost of the development in some cases. Since 2008 this has been restricted to the 60–75 per cent range. Lending rates are considerably above bank base rate, often 200–500 points above base (i.e. if the Bank of England base rate was at 5 per cent, lending on development projects would be at 7–9.5 per cent). The lenders will look at both the projected costs and the Gross Development Value (GDV) in determining the loan. Note if the lender is willing to fund at, say, 60 per cent of the GDV, this may cover 100 per cent of the development costs.
- *Merchant banks* – Merchant banks do not normally fund schemes directly themselves but for larger developers and larger schemes bring together consortiums from their investor clients to provide development funds. They can also sell debenture (debt instruments) for their larger corporate clients, either privately or via the stock markets (see below).
- *Stock market debt* – Large developers and property companies with stock market listings can raise debt finance via the stock market by selling tradable loan stock that pay an annual coupon (interest) rate and whose purchase sum is returned to the buyer at the end of the term. These debentures are normally arranged via merchant banks. They can also sometimes be convertible to shares (i.e. equity) at points during their lifetime. These tend to be the cheapest ways of obtaining finance though only available to the largest companies with good track records.
- *Specialist lenders* – There are a number of specialist lenders in the sector who only lend on property investment and development projects. They tend to operate at the lower ends of the funding market; although many have a lower threshold of £500,000 some will fund at levels lower than this. The rates tend to be higher than the high street banks.

The debt can fall into different classifications.

Senior debt loan

A senior debt loan usually covers the first part of the development costs (although it can be arranged against gross development value). Interest payments can be deferred and regular drawdowns can be agreed in advance for small projects with the interest 'rolling up' as the project proceeds. Developers would normally pay arrangement fees, monthly interest at bank base rate plus 2 per cent to 4 per cent and in some instances, exit fees.

Mezzanine loan

A mezzanine loan is a second charge loan on top of the senior debt loan, hence the name 'mezzanine'. This is similar to a short-term bridging loan and they are often used to fund the gap between the level which conventional funders will go to and the equity funds (if any) available to the developer. A lender has less security on this loan and therefore they are higher risk and the developer should expect to pay much high interest rates than on the primary loan but can achieve loan to value of as much as 90 per cent to 100 per cent.

Joint venture 100 per cent finance

For larger schemes with great potential it may be possible for the developer to obtain 100 per cent funding on a joint venture basis. The lender will both charge interest on the loan (the interest normally being 'rolled up' and accumulating as it is drawn down and repaid as one lump sum at the end of the scheme) and expect to receive a share of the development profits at the end. There are almost infinite ways in which the profits can be split from a pure side-by-side basis to various horizontal and vertical tranches. (Note other joint venture schemes see the land owner provide the site without other financial consideration as their share of the development costs. These parties usually share in the profits at the end of the scheme.)

Grants and public sources of finance

The final source of funds is from the public purse, either from local, regional or national government level or international level such as from the EU or World Bank. These sources are scarce and difficult to obtain but are often important sources of finance in economically deprived areas. Even in these regions, however, these sources almost never provide all of the development funding. Most of this funding is what is termed 'gap' funding. The funds are used at the minimum level to make a scheme viable. Many developments in these areas have insufficient end value to pay for the site and cover development costs plus give the developer a reasonable profit. The public sector, either directly or through an agency of government tasked with regeneration, will top up the scheme to make it viable. These 'leveraged' funds make the most efficient application of public funds and this model, largely pioneered in the UK, is now widely used elsewhere.

The funding arrangements have a huge impact on the viability of a development project. At an initial appraisal, it is unlikely that the appraiser will know what the funding arrangements will be for the scheme and, therefore, broad-brush assumptions will be required. However, even in these cases, some basic information will be required. These will include:

- Whether equity or debt funding is envisaged.
- Whether lenders are willing to fund the type of scheme proposed.
- An approximation of the interest rate applicable to the type of scheme.
- What additional costs the loan will incur (arrangement fees, periodic costs, monitoring fees, exit fees etc.).

As the scheme progresses and more details are known, the appraiser can build in more sophistication to the feasibility study:

- What loan-to-value or loan-to-costs ratios lenders are seeking and, therefore,
  - How much equity will be required (and what cost is ascribed to this element)?
  - How much, if any, mezzanine finance is needed and where it will come from and at what cost?
- Is gap funding required? If so where will it come from and does the scheme meet the criteria?
- Is any joint venture proposed and, if so, how is this to be structured?

Later in this book we will see how the appraisal can go from very simple financial assumptions to highly complex ones, to the extent that a scheme may be reprogrammed in order to minimise financial costs.

As noted above, this is a vital area that has a significant impact on project viability, and it is an area where more sophisticated appraisal models have a distinct advantage over the simple traditional residual approach.

Different procurement methods and their influence on the development appraisal

The final preliminary consideration is how the scheme is to be procured. By this I am generally referring to the type of building contract but this can extend to the whole of the development project.

In a traditional development there is a separation of function between developer, the design team, who include the architect, quantity surveyor, structural engineer etc. who work directly for the developer, and the building contractor (or contractors) who actually physically build the project. There is usually a close working relationship between the developer and the design team and an adversarial relationship between the design team/developer and the contractor. The impact on the appraisal is that all the costs have to be allocated separately to all elements.

Many in the construction industry have pushed for an alternative form of procurement, the design and build route. Here, the design and construction element is provided by the contractor, with the procurer creating a brief which the contractor and his team fulfil. This type of procurement is usually proposed as being faster (in itself this will reduce costs) and less expensive in terms of the cost of design, though many developers producing commercial investment property dislike the lack of control over the end product. This method is, however, widely used. In the appraisal, a single lump sum can be ascribed to cover both construction and design, though the developer may need their own shadow professional team to ensure that the development's final form is in accordance with the original brief, and this team's costs will have to be accounted for.

The final broad type of appraisal sees this taken further. In Public/Private Partnerships (PPP) and Private Finance Initiative (PFI) projects, a full turnkey service is provided where the building or buildings are produced fully complete and ready to use (i.e. they are fully fitted out) – sometimes this also includes the staffing and operation of the building for a set period of time. These types of projects are, most certainly, developments, but are very specialised and, although broadly appraised using the principles used throughout this book, strictly they are outside the scope of this book.

Note

1   To those asking the question; 'what about the interest charge on the borrowed funds?' the answer is that the charge is the same whether the funds are borrowed or not. With debt, there is a 'real' interest charge, with equity funds there is an equivalent charge due to the opportunity cost lost by investing in the project. Equity should, however, be marginally cheaper because there will probably be an arrangement fee charged on the loan.

7 The financial appraisal of development schemes – investigations required to construct the appraisal

Site and land assembly

All development needs land, a site. Sometimes it is the existence of this site that lead to the development, i.e. that the developer sees a site that is ripe for development and the site's layout, characteristics and location then dictate the form and type of development that will take place. A common example of this is a potential site located in a residential area. A vacant site, or perhaps a large garden with an existing building but with sufficient space to enable additional construction to take place on, will almost certainly be developed for residential use. The type of building will usually be one that is in keeping with the surrounding buildings. Here the development type and size is essentially defined. Similarly a site on an existing industrial estate will almost certainly be developed for new industrial or warehouse buildings even though the use class order see most of these buildings classified as B1. (Note that the Use Class Order for England and Wales is included as Appendix A.)

It is a different picture when a developer has a development in mind – of a certain type or size – and needs to identify a suitable site for a scheme to go ahead. Examples include housebuilders, who quite frequently landbank, finding sites suitable for future development well in advance of a project start (sometimes several years before). The developers of retail warehouses will look for locations usually on the edge of existing town or city centres with good car access, good visibility, low competition, and ripe for gaining retail planning consent. For other types of development, including owner occupied buildings, the site will have to meet the requirements of the business or operation and precise specification defined by the end user.

Sometimes it's possible to identify a single existing site suitable for the project. However, for larger developments, it may be necessary to assemble the site, to add several sites together to make them viable. This is the case with large-scale shopping mall developments and regional shopping centres. These developments may need many years of site assemblage before the scheme can go ahead. In some cases it may require the compulsory purchase powers of the local planning authority to be used. All of this will have to be incorporated into initial appraisal. Land acquisition over many years is difficult to appraise due to the changes in value over time as the market fluctuates. There's also the increasing risk that news of the start of the scheme may leak out which will cause existing land owners to raise their asking prices in the knowledge that their sites are essential for the project.

There's a bit of chicken and egg in this situation, in that to assess whether a scheme on land is viable, actually requires the carrying out of a development feasibility study. The process of land value appraisal is covered later in this book.

Data collection required for a development appraisal

The site appraisal will involve gathering data on the physical characteristics of the site, its location and context and on its planning status and history.

Planning data required

Given the importance of gaining consent for a project, determining the planning situation is vitally important. The UK does not have a zonal planning system. Each site's planning status is decided individually; however this will be in accordance with the Local Development Framework prepared by the local planning authority. This in turn has to be in accordance with the national planning policies and legislation, including the Town and Country Planning Act, the Use Classes Order, the General Development Order (which allows certain types of development to take place without the need for formal consent). Planning decisions are also meant to have regard to the Government's Planning Policy Statements (PPS), a series of guidance notes that planning authorities should have regard to in making their plans and in making planning decisions (note these are to be replaced by a single National Planning Policy Framework). A developer should have regard to all of these documents.

It should be obvious that a site with a consent in place for the type of scheme envisaged will be relatively easy to appraise (though it should be noted that there are two forms of planning consent; outline and detailed. With the former, there is a general indication of the type, size and form of the development allowed, with the latter the actual detailed plans of the building and its surrounds are approved). Where there is no consent (or consent has been refused for the type of scheme proposed) then the appraiser must form a judgement as to the likelihood of consent being obtained, either through the normal planning process or through the appeal system. This judgement will be based on collecting and assessing the following information:

- Use type
- Use history of both the site and neighbouring sites
- History of any objections to the development of the site
- Existing planning consents in place on the site
- Lapsed planning consents
- Planning applications and consents on other sites in the area
- Any documentation on planning restrictions on the site.

Most of this data can only be collected by visiting the local planning authority; however if possible a discussion should be had with a local planning officer to determine the planners' attitude to the proposed use, likely objections to the plans, the planners' goals, if any, for the site and the possibility of over-turning any existing planning restrictions on the site. The appraiser will have to form a judgement as to the time this will take and the costs (e.g. for planning consultants) that will be incurred.

Location

The location of the site is vital. It used to be the case that the only way of determining the qualities of the location of a development site was to visit it and spend some time getting to appreciate the site's qualities and its context. Whilst this is still important and recommended, it is now possible to make 'virtual' site visits using Google Earth and this may save considerable time at the initial appraisal stage.

Essentially, the appraiser will be collecting data on the micro and macro locational factors.

Micro location factors look at the detail of the site itself. They include:

- Ease of access and any restrictions (traffic, pedestrians, potentially conflicting uses) that might affect the construction of the project. A greenfield site in an edge-of-town location will be far easier to develop than one in a city centre. Similarly sites in congested central London will be far more expensive to develop than a site in a smaller city or town. The appraiser should look very closely at this issue.
- A second area to consider is site servicing, particularly water, electricity and gas supply as well as sewerage. Ironically the issues here are often the reverse way round to the accessibility; urban sites are usually well serviced and yet servicing rural sites can be more problematical. The corollary of this is that any works in the ground in a city centre will inevitably encounter services that will have to be either protected or re-routed, again at great expense.

If this is the cost side of things, the micro location can greatly influence values as well, particularly for retail but with significant issues for hotels, offices, residential and, sometimes, industrial buildings as well. A distance of only a few yards in retail can make a huge difference in rents; just by being located in the city centre of Manchester will not be enough to achieve the very top value. A shop would need to acquire premises in King Street, Saint Ann's Square and Market Street to do this. Similarly in London, Oxford Street and Regent Street are renowned as being the best retail locations. Even then the very highest values will be achieved at specific locations on these streets. These are sometimes termed the '100 per cent prime' locations and form a very small proportion of the total stock, even in prime locations, and are derived from the areas with the highest footfall and visibility.

Other property types are location sensitive but not to the extent that retail rents are influenced by this factor. There are micro location factors that influence rents in many office locations: solicitors tend to agglomerate around the law courts, surveyors near to commercial legal practices and vice versa, media companies around TV studios, etc. but the actual location is often relatively footloose. In analysing rents the appraiser must also take this into account.

The macro location is the broad location in a region/market sector/city where the property is located. There are some natural overlapping factors with the micro location but essentially they involve broader issues which the appraiser must examine including those which affect cost:

- Access to the road network which will allow the delivery of materials and labour to the site.
- Characteristics of the wider area that may affect the programme or specification of materials of the project. These include restricted hours of working in densely populated or quiet residential areas and the need to use traditional/highly specified materials in conservation areas or where there are wider requirements such as occur in the UK's National Parks (for example, in the Lake District National Park walls have to built in either natural stone or be in white-painted rendered blockwork and roofs be covered in natural slate). All of this has to accounted for in the construction costs. Similarly construction in cities such as Bath and Edinburgh will have to be in keeping with the materials of the historical townscapes and the building forms – which will affect value – will be restricted.

Topography, vegetation, heritage, etc.

There are a number of site specific attributes that the appraiser must have regard to in connection with these factors. These factors will affect both cost and/or value. An ideal site will be cleared, regular in shape, of sufficient size to easily accommodate the development and have no unusual ground conditions. Having all of these characteristics in place is rare.

Some factors, those on the surface, will be relatively easy for the appraiser to determine, even in a preliminary appraisal. Others, beneath the surface, will only be established from prior experience of the locality or after investigations have been undertaken (which themselves will need to be allowed for in the appraisals).

Factors to consider include:

- *Topography*
  - Size and shape of site.
  - Slopes and gradients of the site.

- *Geology and sub-surface conditions*

  - Ground bearing capacity – will special design considerations have to be made for the foundations of the building? If so this will be at a cost.
  - Hard rock requiring removal and difficult excavation.
  - Water table – if this is high this will make construction more difficult and water resistance measures will have to be incorporated in the design.
  - Connected with this is the site's location relative to earthquake risk zones.
  - A significant issue in European cities is the presence of unexploded bombs from WW2.

- *Flora and fauna*

  - Issues include the presence of trees on site. Mature trees may be subject to a tree preservation order which prevents their removal without consent. This may cause design changes or delays whilst an application is made to overturn the order.
  - Japanese Knotweed can be an issue. It is a prescribed species in the UK and, although digging up the underground rhizomes will eradicate it from the site, the waste produced is classified as being hazardous and is difficult and expensive to dispose of.
  - Rare flora and fauna being discovered on the site (or even suggested) can put a hold on development whilst investigations are undertaken and may even have more significant effects on the scheme.

- *Heritage*

  - There are two big issues in historic European cities; the first we have already mentioned above, the need for the development to be undertaken within an often historical city environment which has a huge impact on materials and designs etc. Some listed buildings will require consent to be changed, others may not be able to be touched at all. Some of these factors may be obvious; others may need to be investigated by consulting the planning authorities and heritage groups.
  - The second issue is related and is to do with archaeology. Most cities in Europe have had long histories and there is always the strong possibility of archaeological investigations being required in most city centres. There is a statutory requirement for this to take place in the UK, which the developer must agree with the planning authorities and pay for. This was initially introduced in Planning Policy Guideline (PPG) 16 in 1990 and has recently been amalgamated into a single statutory instrument in Planning Policy Statement (PPS) 5 (2010) which also allows for constructional details of the original structure to be recorded. The Local Planning authorities have a brief to enforce this area. The appraiser should determine the likelihood of archaeological works being required and factor in the cost, programme delays and

any design implications the remains have for the scheme. It is not uncommon for there to be a requirement to protect remains from further damage or to display key components found within the building structure. Sometimes this can add to the appeal of the development. In York, for example, excavations in Coppergate prior to the development of a shopping centre revealed a significant part of Viking York. A decision was made to open a museum recreating part of the Viking town and the dig and this, the Jorvik Centre, has attracted 20 million visitors since 1984 and the shopping centre has clearly benefited from this.

- *Contamination of sites*
  - This major issue is related to the site's previous use.
  - The legal definition of Land Contamination is found in Part IIA of the Environmental Protection Act (1990). Section 78a (2) of Part IIA of the Environmental Protection Act 1990, defines it as:

    Any area which appears to the local authority to be in such a condition, by reason of substances in, on or under the land that

    a) Significant harm is being caused, or there is a significant possibility of such harm being caused, or

    b) Significant pollution of controlled waters is being, or is likely to be, caused.

  - Although the principle is generally that 'polluter pays' often this is impractical as the contamination has resulted from historical industrial use and the polluter has long since gone. As a result the burden usually falls either on the developer or a body tasked with encouraging development as part of a regeneration programme. As a result, local authorities often use the planning system, rather than Part 2A, to encourage remediation of land affected by contamination. The idea is that: (i) Remediation will often be funded by redevelopment, and the planning system can and should secure appropriate investigation and remediation of land and (ii) Part 2A measures should be held in reserve for use where there is no suitable voluntary solution. Planning Policy Statement 23: Planning and Pollution Control (PPS 23)1 explains the relationship between the two regimes. Essentially, after carrying out a development and commencement of its use, the land should not be capable of being determined as contaminated land under Part 2A.
  - The appraiser should not only be aware of these major sources of contamination but also that smaller scale activities can result in problems. For example, petrol filling stations and dry-cleaners are associated with contaminated sites. In addition contamination may exist within the structure of existing buildings, particularly with regard to materials such as asbestos, a highly hazardous material widely used for fire protection.

–   The Environment Agency estimates that there are 300,000 ha of land that may be contaminated in England and Wales, approximately 2 per cent of the total land area. This may seem to be a low percentage but much of this previously used land is in areas ripe for development, particularly for residential schemes.

–   The cost of remediation can be considerable and the appraiser should approach this area with great caution as it can have significant impact on costs and programming.

Market analysis and appraisal

The majority of development undertaken in most countries is for market-driven development hence it is vitally important to gather data on the market and anything that will affect it. One of the complications of the property market is its complexity however; for each type of property – residential, office, retail etc. – each have their own sub-types and sub-markets. Whilst general trends are useful, data is required on the specific market. In addition, property has the potential for several interests to exist in the same property. A large proportion of residential properties are developed for owner-occupation but the investment market is growing in the UK and some developments are aimed squarely at meeting the demand for this market as well. The commercial property markets are even more complex and the investment market is even more significant. Whilst demand in the rental markets for commercial property usually coincides with the most highly demanded investment properties, a developer has to ensure that the product meets the current requirements from the investment community. Often the funding for a scheme is dependent on this market.

These complexities and the importance of the information means that the appraiser will and should spend considerable amounts of time researching these areas whilst constructing the initial appraisal, and that this research should continue if the project proceeds.

Market trends and analysis

A characteristic of property markets is that there is no central source of information. The appraiser will have to go to many different sources to obtain the required information. There is no substitute for talking to experts who are directly involved in the marketplace. This will provide the most accurate and the most direct information. There are problems with this, not least of which the conversation with the market participants might reveal something about the plans for the development. It may be necessary to appoint a local agent and bear what might be non-recoverable costs if the development ultimately proves to be abortive. If neither of these avenues is open, an appraiser may have to rely on secondary, indirect data, particularly at the early stage of the appraisal.

For residential property there is now a substantial amount of data available online. This includes price data, time to sell, price trends and current market prices for both used and new properties. There are a number of house-selling marketing sites available and the developer has considerable choice and depth of information available.

With commercial property this is much more difficult. There are some free sources of commercial data particularly from the websites of the main property press sources in the UK, *Estates Gazette* and *Property Week*, but this data tends to be limited, outdated and sometimes inaccurate. There are pay sites which allow more comprehensive data and again these are expensive and are rough limited use. Other sources of data include:

- Planning consent obtained from the local planning authority give some idea of the trends in the market place as well as the likely competition.
- Some local governments give information on available space and development and a commentary on the local market.
- There are some commercially produced data and reports on key commercial markets and sectors, some of them free, some of them paid for.
- Consulting firms will also use their research departments to produce data on markets (sector and investment trends) and these are sources of data on the most important local markets. This information does tend to be rather general and somewhat historic mainly because they are aware that the other side of the company provides more detailed information to their clients at a cost.

There are some general trends that also need to be monitored and understood. The developer needs to consider these major market trends in designing the scheme. Examples include offices, which became decentralised in the 1980s and 1990s, and open plan from the 1970s onwards. Markets are still dynamic and future trends in offices seem to be towards low carbon, low energy use and less dependent on car access. The highest value is likely to be achieved in those offices which are close to other transport nodes.

Similarly in retailing, historic trends include the move towards mall shopping, the development of retail warehousing from the 1980s onwards, the continuing rise of the supermarkets, new into the UK in the 1960s and now dominating the retail market. Future trends in retailing are much more difficult to predict; although still a relatively small percentage of the total spend, the internet and online shopping has a huge impact on bricks-and-mortar retailing which retailers in the property market have at the time of writing (2013) not yet fully adjusted to. Retailing on the high streets and in malls globally has never been weaker than it is now. It is likely there will be big changes ahead. Industrial property has seen a move from heavy industry to light industry and from manufacturing generally to distribution warehouses, often serving the super-markets and the new breed of online retailers. It is likely that this trend will continue, but again this sector is heavily reliant on road transport currently and may see substantial changes in the future if fuel costs continue to rise and environmental concerns continue to grow.

It is very hard to be prescriptive about these trends. No one has perfect future sight nor can they predict with any reliability what the future trends in retailing or any other property sector will be. The appraiser must keep abreast of trends and do a degree of looking forward in preparing the feasibility study.

Sales and leasing supply and demand

There is a clear overlap with the preceding section; however, rather than collecting evidence in support of values, this examines the two sides of the demand and supply equation: how many of the type of property proposed are either taken up or demanded over a given time period or at the point of development and how much supply, or units of property, are available in the market, either at the point of development inception, or, and this is probably much more important, at the time that the product is expected to come on the market. To do this the appraiser will have to gather considerable data on the market and also gain a good understanding of the market dynamics.

For residential property, as noted above, there is now a considerable amount of secondary information available online about the property market which should allow an understanding to be gained as to what is selling, how quickly it is selling and how well supplied the market is. However, to gain a real understanding, there is no substitute for actually talking to agents in the market if this is possible.

For commercial property, again there is some secondary data about markets available. This includes reports made by the research departments of the leading commercial property surveyors who will report space available in the market place concerned and may also try to calculate the number of months (or years) supply left in the market based upon the average annual take up of space. There may be also some indication of unrealised requirements in the market. Major space occupiers in all sectors sometimes indicate the specifications, size and locations where they would like to take space in general announcements to the market. Perhaps more often, this information is only passed to the agents they instruct who carry out confidential investigations into the market. This underlines again the preference, if possible, for talking to market participants when appraising a development in order to obtain accurate and reliable data on supply and demand. This is particularly important in interpreting the market data because raw take-up figures can be misleading. There can be substantial unfulfilled requirements in a marketplace even where there are considerable amounts of space on the market. This usually is the case where the existing stock does not meet the requirements of the end user. Many office markets have these characteristics; occupiers are seeking modern, flexible space whilst all that is available is older, inefficient, unattractive buildings that are of no interest to them. This occurs in many markets and many sectors and underlines the importance of doing in-depth research to get a proper understanding of the market into which the development is going to be launched.

It should be noted that in the commercial market some developments are tailored towards meeting the requirements of a known end user. If this end user can be secured then this greatly reduces the risk of the development (and also, usually the cost as financing will fall if there is an end user in place). Having an end user is largely essential in a downturn; a scheme would not be fundable by a debt reliant developer without a substantial prelet being in place. In better markets, many developers prefer the speculative market; this type of development generally produces the highest return and capital appreciation. The type of property produced for this approach has to appeal to widest range of possible end users and, for this sector, the more general type of market data is generally applicable; this would not necessarily be the case for the tailored type of development, which, by definition, is likely to be a more specialised building. The appraiser needs to have a careful appreciation of just what kind of development is proposed before an assessment of the market supply and demand data can be made

Financial and economic trends

Property has no intrinsic value in its own right; the demand for property is a derived demand dependent on other economic activity. This is something that, perhaps, many forget, particularly as the picture is complicated by property also being seen as an investment medium and store of value in its own right, somewhere where people put money when they are retreating from other investment mediums – stocks and shares, bonds and liquid cash in particular. Fundamentally however, all property – commercial and residential – is dependent on wealth created in the underlying economy. A good residential market requires good job security, rising incomes and general confidence about future prospects. Commercial markets need economic growth, job creation, high levels of corporate investment and increasing profitability. Appraisers should gain an appreciation of the key economic trends and monitor the main economic indicators issued by the government, the central bank and even by the RICS. In particular they should monitor employment, inflation and GDP indices. This book is primarily aimed at a UK audience; however just as development has become more of an international activity so global economies have become ever more interlinked – as was self-evident to all in the global financial crisis that commenced in 2007. Developers and appraisers should be as aware of what the US Federal Reserve and the European Central Bank are saying and doing as they are about the Bank of England. In addition it is quite clear what is happening in the property markets of Eire, Spain and Greece (and even the sub-prime mortgage market in the US) impacts via the financial system on every local property market in the UK. Appraisers should not be blinkered only to the local property market information.

There is one financial indicator and instrument that appraisers should have particular regard to – interest rates. They have become the primary tool of economic management by governments, used as the principle instrument of inflation control (this is another reason for monitoring inflation rates and forecasts). Property markets in general and development in particular are sensitive to interest rate movements. The effect is twofold; higher interest rates can affect values either directly via mortgage rates for the residential market (one reason that residential property prices have not fallen as much as they might have during the global financial crisis is that interest rates have remained at a historic low in the UK) or indirectly, with high borrowing costs impacting on commercial profitability. The second effect is directly on cost – increased borrowing costs directly increases the cost of development. It should be noted that general interest rates are only indicative of the rates that may be charged to developers; the attitude of lenders to the sector is more important. The GFC is a case in point; the Bank of England base rate was set at 0.5 per cent from March 2009 and is still at this level at the time of writing (May 2013) yet lending to developers is at relatively high rates of interest and low low-to-value ratios requiring developers to provide substantial equity funds even on those schemes that lenders are willing to fund at all. The reason for this is obvious; the world in general and Europe in particular had struggled to cope with the financial crisis and the general austerity measures, problematical public sector finances and the struggle to restart healthy economic growth made the environment for development very difficult and uncertain. Anyone constructing a development appraisal has to monitor not only the general tone of interest rates but also the attitudes and practice of the lending community.

Determining building dimensions

Once this data has been collected, the combination of information from the site (which will determine what is physically possible to build), the planning regime (the determinants of the legal scope for development), on the location and the market (determining what sort and size of development the market wants), should give a strong indication of what can be built on the site and its approximate dimensions. This should be sufficient for an initial appraisal; later, with the input of the design team and any planning advisors on the scheme in particular, more detail will become available and the design will become more secure.

Determining the timing of the development and its stages and phases

This is another key area that is difficult to accurately assess in an initial feasibility study but which becomes easier as the scheme design and planning progresses. However, due to the nature of property development, there will always be elements of the timing of the project that will be difficult to assess.

Naturally, the size and complexity of the development fundamentally determine how long a development will take but, in addition, factors such as location, working restrictions and ease of access to the site influence the timing. Even a small, simple project will take longer to complete where the site is located in a busy urban area than on a greenfield site.

Fundamentally, the appraiser has to determine a number of key points in time and period durations in order to accurately appraise the scheme.

These include –

1   *The development inception date* – this is essentially the date to which all calculations are tied to. It is the date when the project is assumed to have commenced and is, therefore, the point where all land values and profitability calculations discount back to.

2   *The site purchase date* – (or the dates of deposits on the land, interim and final payments). This is often (but not always) the same date as the project inception date.

3   *The preconstruction period* – this is often referred to as the planning period and it is the time when the following activities take place:

   a.   obtaining planning consents/negotiating with the planning authorities/ going through the planning appeals process
   b.   building design process
   c.   obtaining the consent/agreement of other bodies (heritage consents, consultations etc.)
   d.   public consultation
   e.   obtaining vacant possession (extinguishing leases by negotiation or by legal means, removing any illegal occupiers)
   f.   negotiating or extinguishing rights of way, rights of light, wayleaves etc.
   g.   assembling the development team
   h.   carrying out site surveys and other investigations
   i.   carrying out enabling work such as site decontamination.

This period is very dependent on the nature of the project; it can be relatively quick and simple with small developments on greenfield sites or very protracted on complex, brownfield or city centre projects.

4   *Construction/on-site work* – there can be some overlap with the above dependent on how much access is possible in the lead up to this phase. These elements can include:

   a.   demolition
   b.   final site investigations/archaeological work
   c.   re-design work if required
   d.   enabling work and site clearance
   e.   construction of new build elements
   f.   refurbishment work to retained sections

Note that on long term, complex projects, multiple phases will be required.

5   *Letting and sale period* – this is the period when the properties produced
    will let and/or be sold. This period can (and usually does) overlap the
    construction period. This is the critical period to estimate accurately as it
    occurs after the majority of 'real' costs (i.e. those excluding interest) have
    been expended on the project and it is therefore the period that interest
    costs, which are normally rolled up and added to the sums already
    outstanding, increase most rapidly. This is shown in Figure 7.1.

    It can be seen from the expenditure pattern that the cumulative spend
    slows to a crawl at the end of the construction period but that interest
    charges continue to rise. If letting and/or sales are slow the interest curve
    can, in fact, steepen. This uncontrolled increase in expenditure is the hidden
    trap in development projects that developers face if the market deteriorates
    over the lifespan of the development. It is particularly important in
    commercial investment properties that are essentially unsaleable to long-
    term investors if not fully let.

    The corollary to the crucial nature of this element of the timing of
    the project is it is this one which is the most difficult to accurately assess.
    It is hard to assess the impact a new property will have when introduced
    into a market today, yet the developer/appraiser is attempting to do this
    at a point in the future. With a complex commercial scheme this can
    be 2, 3 or even 5 years or more in the future. Economic forecasting,
    although it is greatly improved, lacks the accuracy to give more than general
    indications of future conditions. This is where the research the developer/
    appraiser does into the workings of the market, how they gain an
    understanding of its dynamics and workings, will greatly assist in the

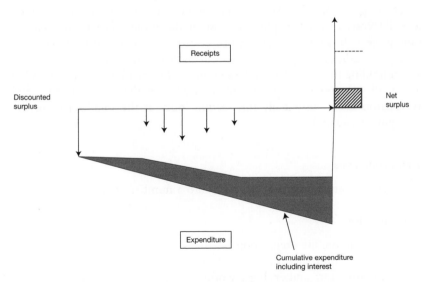

Figure 7.1 Typical expenditure pattern in a development project

determination of this period. It also underscores the importance of the involvement of agents dealing directly in the market place where the development is to be situated is usually vital.

This phase of a development project is always the one that will make or break schemes. It is why little speculative development (i.e. development without a known end user/occupier/purchaser) takes place in an economic downturn; financiers tend to favour schemes with either a complete or substantial pre-letting and/or pre-sale in place as this secures the end of the letting/sales phase.

Whatever, the appraiser/developer should spend more time in trying to determine the length and characteristics of this phase more than any other.

6   *Development completion data or dates* – the final thing, logically, to determine, is the point in the project where the project actually ends. Note that in phased projects, these can be multiple points. Generally, however, this point or points is the date on which the final receipts of the project are received or, in a commercial/investment project, where the final letting takes place and the property transfers from the development project phase to being a standing investment.

Estimating development costs

Build costs

Although the previous sections have touched on factors that have potentially affected cost, many have been general factors that affect value considerations as well. At some point, however, an estimate of cost will have to be made as part of the appraisal. The primary cost will be construction; this will include the build costs themselves plus the cost of the professional team involved in the project. In addition there will be costs to be accounted for which will include any additional capital contributions and fees required by public bodies, costs in holding the site, costs of disposal of the finished building and possibly the overall investment or investments created, plus the cost of finance and any taxation issues. Essentially this section looks at the middle part of the development equation.

Construction costs

Construction costs can be broken down into a number of areas:

* Demolition
    - demolition/site preparation
    - investigation/survey
    - infrastructure and enabling work

- Main build
  - substructure
  - superstructure
  - finishings
- Landscaping (including hard and soft landscaping).

The nature of cost data is similar to other forms of value data. Like any other sort of valuation this is done from collecting comparable evidence. In this case the evidence comes from other construction projects. They have to be fully comparable, the type (the broad type for example, office, retail etc), specification (ranging from basic to mid-range speculative to the bespoke high spec, for example), the location of the buildings (greenfield with its easy access is lower cost than difficult urban sites in major conurbations with restrictions on access) and period of the construction programme (no time restrictions will lead to lower cost whilst for those working to rapid build the time constraints will be much higher).

The appraiser must be very careful about this area; the data must be drawn from similar projects to that proposed. As an example, office construction costs in 2013 can range from around £400/m^2 for a low specification, short life building to £4000–£5000/m^2 for high specification, high status, bespoke projects.

Estimating approaches – superficial area, elemental and detailed breakdown

There are a number of different ways of actually estimating costs. The most commonly used is the superficial error approach. This involves calculating a broad built or projected area and applying a single rate to it. Although this is a quick way of producing appraisal it is relatively inaccurate and will generally only be used in initial appraisal. It can be made more sophisticated by breaking down the project into its component parts, even if this only on a building by building basis. A more accurate method is to take a more detailed elemental cost approach; though the project almost certainly requires a surveyor to be in place for this to be produced, therefore it is usually done as part of a later stage appraisal. This approach breaks down the building into its component parts (substructure, walls, floor, roof, windows etc.) and is much more accurate. The most accurate assessment is the quantity surveyor's approach, which requires a full calculation of all areas, materials and build components that are fully costed. This actually needs a detailed design with full drawings. It is only therefore done at the very final stage appraisal before a project goes ahead.

There are some areas of development work where the quantities are known to a high degree of accuracy. Volume house building is an example of the way the contractors are building repeated examples of similar designs, usually ones that have been built before. It is the closest to the factory style production the property industry gets. The cost of each unit is almost calculated down to the last nail and screw.

Sources of cost information

Sources of data include that drawn from the developer's own experience of previous projects, cost estimates provided by quantity surveyors/construction economist/architects, commercial sources of data such as that produced by the BCIS (the commercial division of the RICS who collect and analyse completed construction projects) and building price books (Wessex, Laxtons and Rawlinson). These provide a range of data sources ranging from broad overall data to highly detailed cost data but the scope is very wide and the appraiser must be able to determine how the project they're considering fits into the spectrum of costs that can be collected.

The BCIS service is the most comprehensive (but also the most expensive) and can provide information that enables the estimation of project costs, price movements and indices and also tools to calculate build duration.

The cost of the professional team

The construction professionals

Professionals involved in development include:

- *Architects* – the scheme designers and (traditionally) the project coordinators, though often today they have been supplanted by professional specialised project managers. In addition to design work the architects also certify stage and/or project completion.
- *Quantity surveyors* – also called construction economists, the QS prepares the Bills of Quantities (if used), cost estimates and cost projections. They also value the work done to date on site, calculating the stage payments due to the building contractor (if a traditional contract, i.e., not design and build, approach is employed). They prepare the final account, agreeing any cost variation with the contractor and sometimes act as project managers.
- *Civil/structural engineers* – they are responsible for the design and calculations related to the structural components of the building. During the project, they also check that the structural work is completed in line with the design and specification, and they certify the work at the end of the project.
- *Mechanical, electrical and services engineers* – these may be required for more complex building to design the services in the building, fulfilling a similar monitoring and certifying role for these areas as the structural engineer does for their areas of responsibility.
- *Project manager* – provides specialist project coordination for the development.

- *Risk, site and project safety managers* – ensure that health and safety requirements of the project are met.

Others may include:

- Landscape architects
- Planning consultants
- Heritage consultants/archaeologists
- Environmental consultants
- Energy consultants.

The cost of these elements can be calculated in a number of ways but should be based upon how these consultants are actually paid. This is normally either as a percentage of the construction cost or a lump sum agreed beforehand or a combination of a lump sum and an incentive fee for meeting a defined target. At an early stage appraisal it is normal to take a percentage of the total build cost estimate ranging from 8 per cent for very simple projects to 14–15 per cent for more complex ones. As a rough rule of thumb, most moderate commercial projects would see professional fees totalling 12 per cent.

One factor already mentioned are Design and Build, Turnkey or PPP/PFI type projects. With this type of building contract, the majority of professional fees are included in the build price and, therefore, no separate allowance has to be made for professional fees (though sometimes the procurer of the project puts their own shadow professional team in place to monitor the project and this cost must be added to the project costs). The appraiser should attempt to assess what contract should be used for the development. In most cases, however, the commercial developer's preference is for traditional contracts due to the level of control over the end product that this affords.

Estimating other development costs – marketing, disposal, statutory costs, road works etc.

Contributions

Contributions tend to result from negotiations with the planning authority. Both sides concede points. The developer can gain the ability to increase the floor space or by gaining permission that allows a use of the site previously prohibited. In exchange, they provide or pay for something that will benefit the wider community represented by the planning authority. Without this arrangement, the local authority would have to pay for these facilities directly. In economic terms it is essentially the internalising of externalities. This is when however it works properly and fairly and is balanced; however, the system is often viewed with suspicion by developers, local politicians and community alike because it can be viewed in different quarters sometimes as being little short of bribery and blackmail. This language is emotive, certainly inaccurate but the view is widely held amongst many involved in this kind of area.

In England and Wales planning gain was allowed under Section 106 of the Town and Country Planning Act 1990. In Scotland the equivalent is a Section 75 planning obligation (Section 75 of the Town and Country Planning (Scotland) Act 1997). The wide criticism led to a report for the government by eminent economist Kate Barker. The report proposed a tax on planning gain, to be known as the Planning Gain Supplement (PGS). PGS was never implemented, and, instead, the government announced in 2007 that a new Community Infrastructure Levy (CIL) was its preferred method of managing contributions from developers. The Government legislated for CIL in the 2008 Planning Act (an act aimed at simplifying the process by which planning consent could be obtained for large infrastructure projects) and the regulations came into force in England and Wales on 6 April 2010. The change of government in May saw little action on CILs; however later in 2010 it was confirmed that the new government would be retaining them albeit in a modified form under the aegis of the Localism Act (2011).

Examples of planning gain include:

- Alterations of road systems, although this may be part of the project anyway.
- Provision of some community building or facility including sports fields and sports facilities.
- Provision of affordable housing etc.

Sometimes the developer is required to actually construct these elements but sometimes the requirement is substituted for a payment, known as a commuted payment.

The appraiser needs to know about the size and timing of these contributions in order to be able to incorporate them in the appraisal. This is often very difficult to do at the initial appraisal stage beyond making an allowance in a contingency percentage or lump sum.

Land holding costs

The major cost of holding land is almost only down to finance. Even when land is bought outright with equity funds (essentially the owner's own money as opposed to borrowed funds) there should still be an interest charge. This is because all funds have an opportunity cost, returns that could have been obtained on the funds if they had not been tied up in the land. To calculate what this interest rate should be, valuers often make a simplifying assumption; the opportunity cost is essentially no different from that which could be obtained if the developer/land owner had sought to lend the money to another developer to do a similar scheme to that proposed. The interest rate on lending is, therefore, the same that is applied to equity. Indeed, a further simplifying assumption that we will encounter when we actually look at the calculation is to assume 100 per cent debt funding for development appraisals.

In practice, almost no developments are funded entirely by debt, and different sources of debt often have quite different costs (being able to deal with these different costs of tranches of funds is one of the great advantages of the proprietary software discussed below). One of those which often in reality has a lower cost of borrowing is lending against land, because the lender will always have security of the land should the borrower default.

There are other holdings costs with land. These include site security, both passive measures, such as hoarding and fencing, and active ones such as guards and CCTV. This is also necessary for public safety and health, to reduce the chances of fly-tipping, gypsy encampments and people injuring themselves on the site itself. The site may also incur taxes, including the Universal Business Rate (UBR) in the UK.

Selling and other costs

The appraiser must include an allowance for the cost of selling and/or leasing the properties and selling on the completed investment if this is to be the case.

Costs may include the following:

- marketing
- preparing brochures and flyers about the project
- press advertisements
- TV advertisements
- marketing suites and sales staff
- agents fees.

Residential agents usually work on a percentage of the sales value achieved, often in the order of 2–3 per cent.

For commercial agents who are tasked with letting commercial space to tenants, the normal rate is around 10 per cent of the initial annual rental value. There may be a requirement for both a local and a national agent in order to capture all potential tenants.

The sale fees for the disposal of the freehold interest (investment) in the properties may be payable to a different agent from the letting agent. The normal fee for investment sale is in the 1–2 per cent of sale price of the investment.

Legal fees

There will be a requirement for a solicitor for any property that is sold as there will be a conveyance of title. Where property is leased there will be fees involved in drawing up the lease. In weaker markets the developer may have to pay the costs of the incoming tenants as well.

Other costs

There are taxes on sale including Stamp Duty Land Tax (otherwise known as stamp duty).

In commercial properties it is necessary to 'Net Down' for incoming commercial costs. These are costs that are borne by the purchaser but which they will adjust their bid to take account of. These costs include the purchaser's own agents fees, legal fees and Stamp Duty Land Tax. These costs are often around 6–7 per cent and the usual way is to deduct this from their gross valuation figure to produce a figure that they should get to give the yield required.

A simple example of this is as follows:

Rental value:	£100,000
Required yield:	6%
Gross value:	£100,000 $\star$ 1/6% = £1,250,000
Incoming purchaser's costs:	6%
Net amount paid:	£1,250,000 / 1.06 = £1,179,245

If 6 per cent is now added to this sum it will produce the gross value of £1,250,000 and the investor will meet their target figure. The appraiser needs to allow for this in the calculations.

Dealing with intangible or uncertain costs

It is almost inevitable at the appraisal stage that some elements will remain either unknown or difficult to estimate at the time

Estimating development values

The previous sections have largely dealt with costs and timings; now we need to turn to the value side of the equation.

Residential projects or elements

For residential development, the appraiser should have collective recent sale values of comparable houses, flats and apartments. They must ensure that they are truly comparable; they have the same type, sophistication and locational qualities. Generally, values are determined based upon whole units of the type of residential property produced (e.g. 1 bedroomed flats, 2 bedroomed house etc.) multiplied by the expected sale price for that type of unit. In some markets, however, the values are determined by unit of floor area.

Commercial property – non-specialised

With mainstream commercial property, the assessment process is more complex as there are two elements to value; the income stream, or rental value, and the capitalisation rate or yield.

We will briefly look at the factors influencing both elements.

Rental value

Rental value is calculated by determining the projected net lettable area and multiplying it by the value that would be achieved per unit of area for that type and class of property. In practice this is determined by the analysis of transactions on comparable properties. This looks simple but is complicated by the number of factors that interact to determine rental value.

In terms of leasing data, to be truly comparable, not only does the letting evidence need to be drawn from similar properties but the terms of leases must be comparable. Notes on the norms of letting structures are given below.

The UK letting market is dominated by long lease terms with periodic periods in which the rents can be reviewed to the market rental value then current. Leases are long complex documents with little standardisation. The benchmark tends to be a vague collection of 'best terms' that tend to be called an 'Institutional lease', i.e. one that is acceptable to institutional standard investors such as pension funds and life assurance companies. These terms have varied over time, for example the average lease term has come down since the property crash of 1990. The terms can be summarised as follows:

- *Length of term* – until the 2007–8 period, the generally accepted term for an 'institutional' lease was 10–15 years, however since then terms on new leases have fallen to well below 10 years on average. This has come down from the 25 years that was traditional for institutions for many years. This length of term guarantees a secure income flow to the investor, particularly valuable when tied to a good quality corporate tenant.
- *Repairs and maintenance* – full repairing and insuring. The tenant is responsible for all aspects of maintaining the property including the structure. The tenant is expected to give the building back in the same order as they acquired it. The landlord will be able to recover all outgoings from the tenant.
- *Rent reviews* – normally every five years to the then open market rental value for similar properties let on similar lease terms, upward only. This clause enables the value of the property to be maintained by periodically adjusting the lease rent. Perhaps the most controversial part of this clause is the 'upward only' component, although strictly this means that the rent cannot fall even if market values have gone down. It does act to maintain the quality of the income flow to the investor.
- *Other terms* – open user clause, open alienation clause, open alteration clause. These clauses are open in that they require the landlord's consent but that consent is 'not to be unreasonably withheld'. This helps to maintain rental values. A recent change is the introduction of authorised guarantee agreements that help maintain the value of the property if the original, good quality tenant leaves.

These lease clauses are amongst the most investor friendly in the whole world, which explains why the UK is popular with overseas investors. They make property investment as clean and as 'hands off' as it possibly can be. The terms may seem onerous but one of the key points to remember is that tenants agree to these terms in the open market. They will only agree to do this due to the qualities of the properties that they wish to occupy.

These long 'clean' leases have always existed in the UK (and Ireland) and have strongly affected UK valuation methods.

The variables affecting rental value therefore include the following.

Lease length

The lease, being the complex legal document under which a property is occupied, has considerable influence on rental values. The length is the most common variable. As a rule the shorter the lease the more a tenant is willing to pay, the reason being that as a lease is a financial commitment, the shorter it is the less the potential cost. Very short leases are, however, often impossible to find takers for whilst a short lease in some circumstances is detrimental to value, particularly in high quality retail areas where retailers want to secure the location for as long as possible.

Lease terms

Generally the more onerous the key lease terms are the less attractive the lease is on the open market and the lower the rent that the tenant is willing to pay. Key terms are the use clause, often used as part of the management structure of retail centres and the repair and maintenance clause.

Physical package

Again the physical package which is being leased is a strong determinant of rental value. The key variables are listed below. It should be noted that these factors interact with each other as well as being individual variables in their own right.

Physical package – size

Generally the rule is that the smaller the property, the higher the overall rental value, indeed the highest rents are often achieved on kiosks in shopping centres. Not only do you have to look at the rate per unit of area but also the rent on an overall basis. Similarly larger buildings often have a quantum reduction per unit of floor area for the size and because the available market for such buildings is relatively small. This effect will vary according to different markets, for example whilst small retail space often produces the highest rent per square unit of floor space, small offices tend not to. The highest rents are being achieved above a certain area then tailing off as the size increases.

Physical package – shape

Again the influence of shape on rental value is determined by the market sector being considered but all tend to see the highest values being produced by regular, unencumbered floor plates, as these tend to maximise the ease of operations. Again retail is perhaps the sector most sensitive to shape, with steps and bends greatly reducing the value.

Physical package – specification

Specification covers a wide range of factors which can range from the standard of finishes to the technical specification provided to allow a business to run. It can include factors such as aesthetics, design and status. Consider that the term 'office' can encompass a single room with bare walls and ceilings, a light fitting and a plug to the Swiss Re 'Gherkin' building in London. They fulfil the scene function but have entirely different specifications. It is generally in the office and residential markets where the specification is the most important factor. With retail, the tenant essentially is supplied only with a rectangular concrete box which is fitted out to the retailer's requirements. Generally a higher specified office or residential property will produce a higher rental value than a similarly located lower specified one.

Physical package – condition

The condition of a building and its fixtures and fitting do have a strong effect on value. The influence is mainly on image and desirability. Although it may be relatively low cost to refurbish a piece of lettable floor space, tenants often do not want the hassle of doing so and thus will discount heavily the rent they are willing to pay.

Location

It may seem surprising that I have placed location down the list of factors that affects rental value. This is, in fact, not to diminish its importance but in fact to stress it. By talking about factors such as size and specification first I can illustrate the variables but also to show how location often supersedes it. The highest values of offices in the UK are often found in the West End of London where the buildings are often old, poorly specified by modern standards and inefficient; the location is paramount however. Notwithstanding this, a modern, technically efficient building with good aesthetics in this location should obtain the very premium rents.

Two additional sub factors to consider are the macro and micro location.

Macro location

The macro location is the broad location in a region/market sector/city where the property is located and which will determine the overall tone of the rent.

Micro location

The micro location is the influence of the specific location on value. In retail the micro location is critical.

Tenant status/size

Many valuers and students of valuations are often surprised when I include this factor in the list of variables that affect rental value, it being drummed into them that it is the premise, specification, location etc. which are the key determinants of value: the tenant is irrelevant. I would acknowledge this to be true with the proviso that in some circumstances the status and size of a tenant strongly influences the rents that are paid. I would cite two opposite polar examples of this: in shopping centres, the 'anchor' tenant, the big operator that attracts shoppers into the centre as a destination is so critical to the success of the development that the rent paid per unit of floor space is heavily discounted and continues to be so through the life of the centre. At the other end of the scale, new small businesses often pay a premium to secure space with landlords.

Market conditions (boom/recession)

The final factor which affects rental value is the state of the market. This will not inherently affect the fundamental tone of the rent. A better specified, better located, better sized and shaped unit will always command the highest rent, but this will affect the rent and the terms and conditions of the deal. Just like any commodity, the price achievable will vary according to conditions. There is no inherent value.

Although this data is essential to determine values the appraiser has to be careful. One particularly important area is in terms of the projection of price rises. An active market with rising property values may not in fact continue into the future. A rising market may well trigger the development but the development itself may cause an oversupply of market. In addition the factors driving demand may change. Demand for property is elastic; changes in economic activity might fluctuate and the demand may evaporate quite rapidly. On the other hand the supply side of the equation is inelastic and other developers may also react to signals from the marketplace and provide supply. This is why the next section is very important.

Yield and capital value

Capitalisation rates are sometimes referred to as yields. They are derived from analysing market evidence, essentially sales of investment properties.

Yields are both the simplest but also one of the most complex areas connected with property valuation. They are simple in that they represent the return on the investment that the investor is either receiving or can expect on an investment. For example if an investment is producing a rental income of £5,000 per year and the owner outlaid £100,000 in total in acquiring the investment then the return is

£5000 / £100,000 x 100/1 = 5%

Similarly the reciprocal of this calculation can be used to find a multiple (known as the year's purchase or YP) which can be used to value income stream of a similar quality. The reciprocal can be calculated by taking the reciprocal of the yield, i.e. 1 / 5% = 20.

If a similar quality income stream was producing £4,500 per annum then this multiple could then be used to find the value:

£4,500 * 20 = £90,000

Fundamentally this is the approach that is taken in valuation and, as can be observed, the process could not be simpler. So where is the complexity?

The complexity with yields arises from two main factors. Firstly yields are subject to similar variables that we reviewed with rents. Secondly there are many different kinds of yields, some of which are just used in analysis, some of which can be used in valuation. This situation is complex and can be confusing. The various types of yield will be discussed below. Initially however, we will discuss the variables that affect yields.

Essentially the variables that influence yield are those which affect the quantity, quality and duration of the income stream. Like the variables affecting rental value, these factors are strongly inter-related.

Sub variables that influence yields

Tenant status

One of the first things that valuers look at when assessing the yield to apply to a property is who the tenant is. It is not the definitive determinant of yield, it cannot be looked at in isolation of variables such as the term and terms of the lease or the quality of the location and/or building but it is highly significant. Effectively, all other variables being equal, a Public Limited company will offer an investor a lower risk than a small, local company so the former should attract a lower yield (and therefore a higher value). The key phrase in that statement is 'all other variables being equal'. To achieve the lowest yields and the highest values the good quality tenant must not be allowed to walk from the lease.

Lease length

Again, all other variables being even, the longer the term the longer the secure income stream is, the lower the risk of interruption and the lower the yield. A lease is a contract that cannot normally be broken, hence a long lease should guarantee an income flow as long as the company stays in business.

Lease terms: break clauses

One way that tenants with superior burgeoning power can introduce flexibility into their occupation and reduce their long-term liability is to introduce so-called 'break' clauses into otherwise long leases. These allow an early termination of the lease allowing the tenant to walk away at fixed points in the lease and subject to following set procedures. Sometimes a penalty is payable. The impact on yield of these clauses is normally negative, i.e. they increase the risk and thus the yield lowering the value. Only in a buoyant, rising market might these clauses have a positive impact on yield.

Lease – other significant terms

The other terms that have a significant impact on yield are very similar to those which impact on rent. Onerous rent review, repairing or user clauses would impact on the future growth potential of the asset. Low growth means than an investor must be compensated by a higher return elsewhere, which can only really come from a higher initial income yield and thus lower value.

Physical properties – nature and quality of the building

Similar arguments are affected by the physical make-up of the building. A highly specified, modern office building with large floor plates will produce lower yields than older office stock because its rental growth potential is higher. There is also the issue of depreciation. There is a fundamental yield difference between, say, industrial, office and retail property with the former having high rates of depreciation (industrial due to wear and tear, offices due to changes in specificational requirements) and the latter, retail, hardly depreciating at all. Industrial buildings will also tend to have higher yields than the other asset classes.

Location

Similarly the location has a very strong influence on yield. The best locations naturally attract the best tenants, developers and investors can generally negotiate long leases and the growth potential of these locations is higher, hence the yields tend to be lower.

This complexity must be taken into account when yield evidence is collected, evaluated and then applied in the appraisal.

Commercial property – specialised

The vast majority of development projects will be for the most common types of property – residential houses and flats/apartments and commercial offices, shops, industrial buildings and warehouses – and, for all these types, the above sections are applicable. However, some developments will be for more specialised types of property. These properties sometimes create problems for the developer/appraiser when calculating construction costs but usually create problems when it comes to assessing value/end worth.

These types of properties include:

- Hotels
- Restaurants
- Bowling alleys
- Bingo halls
- Cinemas
- Other leisure activities such as laser quest etc.
- Indoor ski centres
- Spas
- Health and fitness centres
- Tennis centres
- Football centres
- Golf courses.

The value of all of these types of properties is based upon their underlying business earnings and profitability, i.e. the valuations are more of the business and only partly connected with the physical building and connected physical features created. This creates problems for the developer/appraiser who lacks experience in the sectors; market evidence is far more limited to value them reliably from similar comparables and the appraiser is unlikely to have the in-depth knowledge to determine what long-term income and running expenses, both of which need to be calculated, are reasonable for the business, the building/entity produced and the location.

If faced with such a situation, a 'mainstream' appraiser should seek out specialists in the appropriate field to assist them with constructing the feasibility study; however, the two market leaders in the field of proprietary models, Argus and Estate Master, both have modules that allow the appraisal of properties of this type. An example of the appraisal of an 'operated asset' – in this case a golf course – is included in the Argus Developer section.

Conclusions

From the previous section it should be clear that a development appraisal involves considerable research. There is a tendency to rush to calculation and one of the reasons for this is that the proprietary systems used in the worked examples are often too easy to use and give the impression of accuracy from the superficial quality of the figures produced – but this should be resisted. Even with initial appraisals where the data is limited and the scheme details almost certainly sketchy, as much information as possible should be collected to produce as refined a figure as possible.

In the next section we will look at how the data is put together to create the final appraisal. Essentially we are looking from this point on at the mechanics of producing the appraisals.

8 The mechanics of constructing development financial feasibility studies

An overview of the development of practice – from tables to proprietary software systems

There used to be an argument in development appraisal as to which is the most appropriate model for doing development appraisal – the traditional residual approach or the cash flow approach. In many ways this argument has already been won; most professional appraisers and developers use one of the proprietary appraisal systems and all of these models use a cash flow approach as the calculation engine. It is worth looking at all of the approaches however for some practitioners still adhere to the residual approach, even when using spreadsheet software such as MS Excel, and there are some distinct differences in the cash flow approaches that people are often unaware of. In addition, the residual approach has distinct advantages in terms of summarising the results of a development appraisal; indeed all of the proprietary systems have a residual layout as one of the outputs of the calculation.

The three different approaches that can be used for either profit calculation or land residual calculations are:

- traditional residual approach
- the residual (accumulative) cash flow
- the discounted cash flow (DCF).

The residual approach is essentially a throwback to an earlier time when it was technically difficult to do complex, extended calculations. The residual layout uses lump sums to account for costs and values, and the effect of time is allowed for by a crude calculation of accrued interest.

What is laid out below is a traditional residual model for a mixed retail and office scheme.

Pre-construction	6 months
Construction	9 months
Letting	9 months
Discount rate	10%

Value of scheme

	m^2	Rental/m^2	Income	
Retail net area	212.5	£200	£42,500	
Office net area	830	£175	£145,250	
Total income			£187,750	
Yield		7%	14.29	
Capital value				£2,682,143

Cost of scheme

Building costs					
Retail	250m²	£500	£125,000		
Office	1000m²	£800	£800,000		
Road/site works	250m²	£450	£112,500	£1,037,500	
Professional fees	12.50%			£129,688	
Contingency	5%			£58,359	

Development finance

Construction and planning	10%	half balance £612,773	£77,531	
Letting void	10%	£1,303,078	£96,557	
Letting and legal fees	15%		£28,163	
Sale fees	2%		£53,643	
Advertising/ marketing			£30,000	
Developer's profit on cost	20%	£1,511,441	£302,288	£1,813,729
SITE VALUE IN	24 months			£868,414
Present value of £1				0.826
				£717,697
Less acquisition costs	5%			£35,885
SITE VALUE TODAY				**£681,813**

This is a fairly typical, simple appraisal. Let us now analyse the appraisal to see how it is constructed, piece by piece.

Pre-construction	6 months
Construction	9 months
Letting	9 months

1 *Timing and Phasing*
   This is a relatively typical commercial project. The assumption is the project starts on the day the appraisal is carried out, has a six month run-in period to the start of construction, which lasts 9 months and then is expected to take nine months to let, after which a sale of the investment interest is assumed, after which the development is assumed to be over. In the more sophisticated cash flow approaches, this structure allows different

costs to be allocated to each section and more detailed assumptions to be made about how these costs and receipts occur. Within the rather crude restrictions of the residual model only broad-brush assumptions can be made about these factors, and this impacts on the calculation of the interest charges (see below).

Discount rate   10%

2 *Discount rate*
Strictly speaking, in this calculation, this is an interest rate as it is used to estimate the total interest charge the project accrues over time, but as most traditional residual calculations assume 100 per cent debt financing, it is the common rate that is applied to debt, equity and other sources of finance and is used to discount back the future, accumulated sums, back to the present. How this is done is discussed below.

	m^2	Rental/m^2	Income
Retail net area	212.5	£200	£42,500
Office net area	830	£175	£145,250

3 *Value on completion (a) Rental value*
The top part of the residual equation calculates the amount the developer will receive on completion of the scheme. As this is a commercial, income-producing scheme, the first step is to calculate the rental value. In this case the building has retail on the ground floor with offices above and the two elements have been valued separately. In both cases the net lettable area has been used (the useable space in accordance with the RICS Code of Measurement Practice) though retail space is also frequently zoned in practice (its area expressed as a proportion of Zone A using 6 metre deep zones from the front). The rental evidence for both types has been derived from recent market evidence from similar properties in the locality.

Yield   7%

4 *Value on completion (b) Capitalisation rate/yield*
This also has been derived from market evidence from the recent sales of similar investment properties in the open market, taking the net income/price paid x 100/1. In this case, a compound yield has been used applicable to all elements of the income. Often it is more accurate to use a separate yield for each element as (a) office yields are often higher than retail yields and (b) the investments are usually separable.

Total income		£187,750	
Yield	7%	14.29	
Capital value			£2,682,143

5 *Value on completion (c) Capital valuation*
The capital sum is calculated by multiplying the total income by 1/yield, the reciprocal here working out to a capitalisation rate (also known as a

year's purchase) of 14.285, rounded to 14.29. Strictly speaking, this sum should be netted down for incoming purchaser's costs (see section on yields, above).

Building costs

Retail	250m²	£500	£125,000	
Office	1000m²	£800	£800,000	
Road/site works	250m²	£450	£112,500	£1,037,500

6 *Construction costs*
For increased accuracy, the construction cost calculation has been split into three elements: retail, office and road/site works, the latter of which also includes demolition. This reflects that the elements have very different cost characteristics. Offices tend to be completed fitted out for users to take immediate occupation and therefore come complete with lighting, heating, air conditioning units etc., whilst retail units are normally finished in shell format as individual retailers seek to 'brand' their occupation with their fitting out. Similarly, separating out the external works allows them to be priced more accurately. At a very early stage appraisal, however, even this degree of split may not be possible and an overall lump sum applied. Here the building costs have been derived from similar projects in the locality.

Professional fees   12.50%   £129,688

7 *Professional fees*
An overall percentage figure for professional fees based upon a rough rule of thumb from other projects.

Contingency   5%   £58,359

8 *A contingency sum*
An allowance of 5 per cent of the build cost has been applied to cover unknown elements in the construction.

Development finance

Construction and planning	10%	half balance	£612,773	£77,531
Letting void	10%		£1,303,078	£96,557

9 *Development finance*
There are a number of points in a residual appraisal where interest calculations are made. These areas are important in the calculation as it is where the effect of a key component – time – is accounted for in the feasibility study. In many developments this cost will be 'real' i.e. there will be a loan with an interest charge against it. The interest charge will be dependent on the amount of the funds drawn down and the time taken to repay them, and this is indeed what is being estimated here.

Even when equity funds are used, however, it is still important to use these calculations to accurately determine the viability of a project. All funds used in a project have an opportunity cost, whatever their source. The opportunity cost of borrowed funds is indicated by the interest rate charged. The opportunity cost for equity funds should be determined from the market, derived from similar projects with similar risk profiles. This sounds complex but it is generally accepted that a good proxy for the opportunity cost is, simply, the interest rate that a lender would require to lend against the project. It is for this reason that most appraisals assume 100 per cent debt funding. Later, more in-depth appraisals can and do allocate different costs to different funding sources (this will be covered later in the book) but here the 100 per cent funding assumption is made.

The effect of borrowing is accounted for by calculating the interest which will accrue on half the construction costs for the entire development and planning period of 15 months at the loan rate of 10 per cent. The half balance is used because that is assumed to be the average amount owed to the financiers. This is a crude and inaccurate simplification and represents one of the model's main flaws.

A second calculation of interest is then made on the whole construction costs (which includes professional fees) and the interest assumed to have accrued up to the end of the construction period. This is calculated over the assumed letting up period of 9 months. It should be noted that the interest charge in this period is high; this is a particularly sensitive part of a development; although the building is completed the value will be heavily discounted until a tenant is secured (or a sale made as appropriate to the type of development).

These two interest charges are accumulative, i.e. the calculation estimates the values and costs at the final completion of the development (the final letting and sale as opposed to the physical completion of the building). To calculate the value of the land value (or profit) today this has to be discounted back to the present (in terms of a land value calculation this will account for financing the purchase of the land over the entire period of the development, for a profitability calculation it accounts for the time value of the profit sum received). For a land value calculation the cost of acquisition (stamp duty land tax and legal and surveyors fees) also have to be accounted for.

Letting and legal fees	15%	£28,163

10   *Letting fees*
Letting fees are calculated as a percentage of the initial annual rental value. Normally in the UK this is at 10 per cent but here this sum also includes 5 per cent for legal costs in preparing the leases.

Sale fees	2%	£53,643

11   *Sale fees*
These are the direct cost to the developer (as opposed to incoming purchaser's costs) and are made of 1 per cent fees to the investment/sale

agent and 1 per cent to a solicitor for the conveyancing work. These are common benchmark figures.

Advertising/Marketing	£30,000

12 *Marketing*
Separate sums are allowed to cover the preparation of brochures and press adverts.

Developer's profit on cost	20%	£1,511,441	£302,288	£1,813,729

13 *Developer's profit figure*
As this calculation is to calculate the land bid, an allowance has to be included to cover a reasonable profit for the developer. Here it has been calculated at 20 per cent of total development costs.

SITE VALUE IN	24 months	£868,414
Present value of £1		0.826
		£717,697
Less acquisition costs	5%	£35,885
SITE VALUE TODAY		£681,813

14 *Residual value*
The difference between the sums expected to be received and the total costs expended is the value of the site in 24 months' time. To calculate the sum that should be paid now, this sum needs to be discounted back to the present using $(1 + i)^{-n}$, i.e. 1.10^{-2}, which will account for the cost of interest accruing on the land over the 2 year period. To this should be deducted the cost of purchasing the land (5 per cent). (Note, a more accurate way of calculating this is to solve the equation $L = (L + 5\%) + (1 + i)^{-n}$ where L = Land Value.)

This is the traditional way of doing residual calculations. This approach is very much a simplified calculation which enabled valuers to produce them manually using valuation tables and calculators. Even though this approach was down to the technical restrictions on valuers which no longer exist, this approach has persisted into the computer age, much in the same way as the QWERTY keyboard has, even though the rationale behind its design has gone (the need to avoid mechanical clashing of keys in the case of the keyboard). It has distinct disadvantages; it is so crude that the results are almost certain to be inaccurate and it also cannot calculate benchmark return figures such as Internal Rate of Return (IRR).

As noted, the results the method produces and the calculations shown are more easily interpretable in reports and output when using this residual framework. It is in summarising development projects that the residual approach probably has its main contemporary applicability.

Moving the model on: cash flows and spreadsheets

It will probably come as no surprise that the simple residual model presented above was actually calculated using a spreadsheet. The formulas used in the calculation are displayed in Figure 8.1.

I have split the appraisal into three to illustrate the fact that this is a simplified cash flow. The top part of the equation is a simple investment valuation. The income stream is calculated from the net areas expected to be developed and this income is then calculated using an appropriate All Risks Yield (ARY). Given that the yields on retail and offices are often different (retail generally being lower due, in part, to the lower risk of depreciation) I could have used a split yield approach here and calculated the investment value for each component separately but I chose to use a compound yield as I expected the investment to be sold as a single package. The top part of the equation is shown in Figure 8.2.

The central part of the calculation is the area where the bulk of the expenditure takes place and is also the main area where the interest charges

◢	A	B	C	D	E	F
1						
2	Preconstruction	6 Months				
3	Construction	9 Months				
4	Letting	9 Months				
5	Discount Rate	10%				
6	**Value of Scheme**					
7			m²	£		
8	Retail Net Area		212.5	£ 200	=D8*C8	
9	Office Net Area		830	£ 175	=D9*C9	
10	Total Income				=SUM(E8:E9)	
11	Yield			7%	=(1/D11)	=E11*E10
12	Capital Value					
13						

Figure 8.1 Calculation of value on completion

◢	A	B	C	D	E	F
13						
14	**Cost of Scheme**					
15	*Building costs*					
16	Retail	250	£ 500	=C16*B16		
17	Office	1000	£ 800	=C17*B17		
18	Road/Site works	250	£ 450	=C18*B18	=SUM(D16:D18)	
19						
20	*Professional fees*	12.50%			=E18*B20	
21						
22	*Contingency*	5%			=B22*(E18+E20)	
23						
24	*Development Finance*					
25	Construction and Planning	10%	half bal	SUM(E18:E22)/25)^((B2+B3)/12))-1)*D25		
26	Letting void	10%		=SUM(E18:E25+B26)^(B4/12)-1)*D26		
27						

Figure 8.2 Calculation of construction costs, professional fees and finance charges

◢	A	B	C	D	E	F	G
28	*Letting and Legal Fees*	15%			=B28*E10		
29	*Sale Fees*	2%			=F11*B29		
30	*Advertising/Marketing*				£ 30,000		
31							
32	*Developers Profit on Cost*	20%		=SUM(E18:E30	=D32*B32	=SUM(E18:E32)	
33	*SITE VALUE IN*	24 mnths				=F11-F32	
34	*Present Value of £1*					=(1+B5)^(B33/12)	
35						=F34*F33	
36	*Less acquisition costs*	5%				=F35*B36	
37	*SITE VALUE TODAY*					=F35-F36	
38							

Figure 8.3 Calculation of ancillary costs and present value of land

are calculated. In the residual an assumption is made that the average sum owed on the drawdown of the construction costs and professional fees is half the total cost of these elements for the whole planning and construction period. A second finance charge is then calculated on all of the net balance from the end of construction to the end of the assumed letting void. Both are crude approximations of the actual expenditure and interest accrual.

In the final section, both the marketing, letting and sale fees are assumed to take place right at the end of the scheme and thus accrue no interest. Again these are gross oversimplifications.

The end adjustment shows the calculated future land value discounted back to the present using the finance rate (generally finance rates are used in discounting in feasibility studies as they are taken to represent the opportunity cost of investing money in a scheme), and then adjusted down to take into account the cost of acquiring the land (stamp duty and fees). See Figure 8.3.

It is natural to do these calculations in Excel because, in the same vehicle, you have both the calculation engine and the method of presentation, transferring and printing them off in MS Word just as was done here. However, as noted above, this is not a proper cash flow, it is merely an automation of the traditional residual model. The result is a rather inadequate appraisal. To do things more accurately more consideration must be given to both the quantum and the timing of expenditures and receipts, i.e. they must be placed into an appropriate time framework; a proper cash flow projection.

Residual (accumulative) cash flows and discounted cash flows

Both the DCF (see below) and residual cash flow models employ the same basic approach. They both require a monthly (usually, although other time periods can be used) projection of income and expenditure over the entire time period of the development. The differences are that the accumulative cash flow works forward, calculating the estimated interest that accrues each month and then rolling forward this amount, accumulating the interest and adding it to the 'hard' expenditures (and deducting any receipts) until the end of the project. When the final sale is made (or the project ends in some other way such as being transferred to being a live investment when it is 100 per cent or substantially let), then the project, if viable, will be left with a surplus, either to fund the purchase of the land or as being the future development profit figure. In the DCF, in contrast, the net cash flow from each month is discounted back to the present using either the finance rate or a target rate of return; at the finance rate the resultant Net Present Value (NPV) will be either the Land Bid estimate or the NPV of the projected profit, dependent on the type of calculation being done, or at the target rate of return it tells the appraiser whether the development meets the required performance parameters. Both will be illustrated below, however some of the technical objections against cash flows should be discussed first.

There are problems with cash flow feasibility studies which has meant that some sections of the industry have resisted their use. The first principal complaint about the models is that they are time consuming to produce from scratch, particularly compared with residual models. Once a template model has been developed however, this time can be reduced as the model can be reused for different projects; however there are inherent risks of error when existing models are adapted to meet the needs of different feasibility studies. The second major complaint is the level of detail required in the assumptions that go into the construction of the cash flow template. Again there are elements of truth in this; the models are far more complex and transparent yet many of the assumptions required can reasonably be made from past experience from similar projects and are not that far removed from the sweeping, broad-brush ones made in the residual models.

The final major complaint is one that is valid; the increased complexity of the cash flow models means that there is more risk of error creeping in. These are not so much errors of assumption but more simple mistakes in calculation or cell reference. Spreadsheet cash flow models have to be very carefully audited and, often, there is insufficient time to do this. This is undeniable and is why I am an advocate of using proprietary models where these types of errors can be virtually eliminated. This will be discussed more in Part Two of this book.

Before we turn to these we need to look at the mechanics of the calculations, using the same basic project as we looked at in the residual example (Figure 8.4).

Once the preliminary judgements have been made, the next step is to create the cash flow. The cash flow is simply a time framework representing a forecast of when the expenditure and receipts will occur. For ease of viewing this is displayed in three sections; the preconstruction phase (Figure 8.5), the construction phase (Figure 8.6) and the letting/sale period (Figures 8.7 and 8.8).

The major advantage of the cash flow approach is the ability to much more accurately assess the financing costs of the scheme. This is possible because the cash flow projection gives a month-by-month forecast of the monthly financial expenditure on the scheme; this in turn allows the expected amount of interest to be calculated. Granted this requires the cash flow itself to be a reasonably

Development Assumptions

Finance	
Interest Rate (Annual)	6.75%
Interest Rate (Monthly)	0.56%
Arrangement Fee	£25,000

Timescales	
Planning/preconstruction	6 Months
Construction	9 Months
Letting/Sale	6 Months

Land	
Acquisition Cost	£680,000.00

Areas	Gross	Net	Construction Costs	Rental Values	Yields	Rent Free
Office	1,300 m²	1,079 m²	£800	£185	6.50%	6 Months
Retail Warehouse	1,000 m²	950 m²	£600	£230	6.25%	6 Months
Landscaping (hard & soft)	3,000 m²	3,000 m²	£250			

Fees		Contingency	5%		
Architect	4.00%	Landscape Architect	2.00%	Land Legal	1.00%
Quantity Surveyor	2.00%	Planning Consultant	£12,000	Letting Agent	10.00%
Structural Engineer	1.00%	Planning Fee	£15,000	Letting Legal	5.00%
M&E Engineer	1.00%	Sale Agent	1.00%	Stamp Duty	4.00%
				Incoming purchasers Costs	6.00%
Project Manager	1.00%	Sale Legal	1.00%		
Site Safety	0.75%	Land Agent	1.00%	Marketing	£25,000

Figure 8.4 Base project development assumptions

Months	0	1	2	3	4	5	6
Costs/Values							
Rental Income - Office							
Rent Free - Office							
Sale - Office							
Rental Income - Retail							
Rent Free - Retail							
Sale - Retail							
Purchasers Costs							
Land acquisition	-£ 680,000						
Stamp duty	-£ 27,200						
Construction - Office							
Construction - Retail							
Construction - Landscape							
Contingency							
Architect							-£ 32,800
Quantity Surveyor							-£ 11,950
Structural Engineer							-£ 4,100
M&E Engineer							-£ 4,100
Project Manager							
Site Safety							
Landscape Architect							
Planning Consultant					-£ 12,000		
Planning Fee				-£ 15,000			
Sale Agent							
Sale Legal							
Land Agent	-£ 6,800						
Land Legal	-£ 6,800						
Letting Agent							
Letting Legal							
Loan arrangement fee	-£ 25,000						
Marketing							
Monthly Totals	-£ 745,800	£ -	£ -	-£ 15,000	-£ 12,000	£ -	-£ 52,950
Running Total	-£ 745,800	-£ 745,800	-£ 749,995	-£ 769,214	-£ 785,456	-£ 789,807	-£ 847,200
Interest Charge	£ -	-£ 4,195	-£ 4,219	-£ 4,242	-£ 4,351	-£ 4,443	-£ 4,468
Cumulative Totals (Including interest)	-£ 745,800	-£ 749,995	-£ 754,214	-£ 773,456	-£ 789,807	-£ 794,250	-£ 851,667

Figure 8.5 Pre-construction phase of cash flow

accurate assessment of what will happen in practice; however quite a large proportion of a project can be reasonably forecast – the amount of time it will take to design the project and get the necessary consents, how long it will take to build and how much it will cost etc. Only in very complex projects or ones where elements are unknown (an example of this is refurbishment projects of historical buildings where substantial investigation work may be required during the project itself) will this forecasting prove really difficult. There are some elements even in simple projects that may be harder to estimate – the time a building will take to let or to sell for example – but even here a reasonable forecast is possible. Scrutiny of the cash flow will see that I have adopted the classic S-shaped curve for construction expenditure as well as trying to model more accurately when the construction professionals are typically paid (normally a substantial payment at the time of the placing of the construction contract followed by periodic stage payments) and also elements like the expenditure of the marketing budget.

Whatever, we can come up with a forecast of the monthly balance on a project that allows a much better calculation of the financing cost than with the residual approach. In the latter, the approach is to merely take half the balance of the construction budget plus the professional fees and use that as a proxy for the average drawdown. This is normally a very approximate estimate of the interest accruing on the project.

It is at this point that there is a variance in approaches between the calculation of the residual cash flow and the pure DCF approach. Both use the calculated monthly expenditure forecast, however the residual cash flow requires the calculation of the cumulative expenditure as well (note in the cash flow examples above, the residual (accumulative) cash flow method is shown).

7	8	9	10	11	12	13	14	15
-£ 104,000	-£ 156,000	-£ 260,000	-£ 260,000	-£ 156,000	-£ 104,000			
				-£ 90,000	-£ 150,000	-£ 180,000	-£ 90,000	-£ 90,000
						-£ 187,500	-£ 375,000	-£ 187,500
-£ 5,200	-£ 7,800	-£ 13,000	-£ 13,000	-£ 12,300	-£ 12,700	-£ 18,375	-£ 23,250	-£ 13,875
				-£ 16,400				
				-£ 23,900				
				-£ 8,200				
				-£ 8,200				
-£ 1,040	-£ 1,560	-£ 2,600	-£ 2,600	-£ 2,460	-£ 2,540	-£ 1,800	-£ 900	-£ 900
-£ 780	-£ 1,170	-£ 1,950	-£ 1,950	-£ 1,845	-£ 1,905	-£ 1,350	-£ 675	-£ 675
						-£ 3,750	-£ 7,500	-£ 3,750
								-£ 12,500
-£ 111,020	-£ 166,530	-£ 277,550	-£ 277,550	-£ 319,305	-£ 271,145	-£ 392,775	-£ 497,325	-£ 309,200
-£ 962,687	-£1,134,008	-£1,417,000	-£1,700,959	-£2,028,271	-£2,309,029	-£2,713,267	-£3,223,645	-£3,548,180
-£ 4,791	-£ 5,442	-£ 6,409	-£ 8,007	-£ 9,613	-£ 11,463	-£ 13,053	-£ 15,336	-£ 18,219
-£ 967,478	-£1,139,450	-£1,423,409	-£1,708,966	-£2,037,884	-£2,320,492	-£2,726,320	-£3,238,980	-£3,566,400

Figure 8.6 Construction phase of cash flow

16	17	18	19	20	21	22	23	24
								-£ 99,808
								£3,071,000
								-£ 109,250
								£3,496,000
								-£ 394,020
-£ 16,400								
-£ 11,950								
-£ 4,100								
-£ 4,100								
								-£ 65,670
								-£ 65,670
								-£ 41,812
								-£ 20,906
-£ 12,500								
-£ 49,050	£ -	£ -	£ -	£ -	£ -	£ -	£ -	£5,769,865
-£3,615,450	-£3,635,511	-£3,655,960	-£3,676,525	-£3,697,206	-£3,718,002	-£3,738,916	-£3,759,948	£1,988,768
-£ 20,061	-£ 20,450	-£ 20,565	-£ 20,680	-£ 20,797	-£ 20,914	-£ 21,031	-£ 21,150	-£ 21,269
-£3,635,511	-£3,655,960	-£3,676,525	-£3,697,206	-£3,718,002	-£3,738,916	-£3,759,948	-£3,781,097	£1,967,499

Figure 8.7 Post–construction phase of project

	Gross	Discounted
Profit	£1,967,499	£1,719,686
Return on Cost	47.06%	41.14%
Return on Development Value	31.87%	27.86%
Development Yield	10.00%	10.00%
IRR (monthly)	3.323%	3.323%
IRR (annual)	39.88%	39.88%
Rent Cover	56.47 Months	49.36 Months

Figure 8.8 Performance analysis

The residual (accumulative) cash flow calculation

Both the traditional residual approach and the residual (accumulative) cash flow use the same principles (Figure 8.9) in that the interest charge is estimated and added to the cost of development. At the end of the development period this total, rolled-up cost is deducted from the sale receipts and the surplus is discounted back to the present. With land valuation, the discounting is to account for the cost of land holding, with the profitability assessment (as here), the appraisal is calculating the present worth of the profit. With the cash flow approach the interest charge estimate should be more accurate than with the crude traditional residual approach.

The procedure for calculation involves calculating the balance owed at the start of the month (which except for month 1 will include interest as well as physical expenditure), adding to this the interest accruing on this balance at the monthly interest rate and then adding the construction expenditure that has occurred during the month (a simplifying assumption here is that any expenditure during the month occurs at the end of the period thus not attracting interest until the end of the following month). The bottom part of the cash flow is presented in Figures 8.10 to 8.13.

With this approach it is possible to keep a cumulative total of expenditure, principal and interest combined. The final figure in the cumulative row, the £845,767 in month 24, is the projected profit figure at the end of the development. Because it is positive we know that it is profit. Strictly this is a future value and it should therefore be discounted back to the present at the

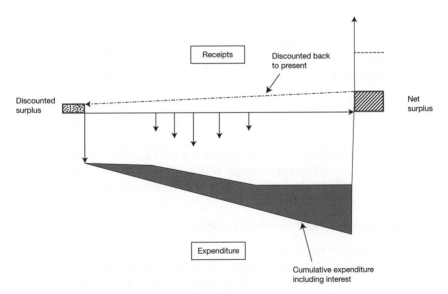

Figure 8.9 Principles involved in traditional residual and residual (accumulative) cash flow approaches

Monthly Totals	-£	745,800	£	-	£	-	-£	15,000	-£	12,000	£	-	-£	52,950
Running Total	-£	745,800	-£	745,800	-£	749,995	-£	769,214	-£	785,456	-£	789,807	-£	847,200
Interest Charge	£	-	-£	4,195	-£	4,219	-£	4,242	-£	4,351	-£	4,443	-£	4,468
Cumulative Totals (Including interest)	-£	745,800	-£	749,995	-£	754,214	-£	773,456	-£	789,807	-£	794,250	-£	851,667

Figure 8.10 Extract from cash flow: land purchase and pre-construction/planning stage

-£	111,020	-£	166,530	-£	277,550	-£	277,550	-£	319,305	-£	271,145	-£	392,775	-£	497,325	-£	309,200
-£	962,687	-£1,134,008		-£1,417,000		-£1,700,959		-£2,028,271		-£2,309,029		-£2,713,267		-£3,223,645		-£3,548,180
-£	4,791	-£	5,442	-£	6,409	-£	8,007	-£	9,613	-£	11,463	-£	13,053	-£	15,336	-£	18,219
-£	967,478	-£1,139,450		-£1,423,409		-£1,708,966		-£2,037,884		-£2,320,492		-£2,726,320		-£3,238,980		-£3,566,400

Figure 8.11 Extract from cash flow: construction stage

-£	49,050	£	-	£	-	£	-	£	-	£	-	£	-	£	-	£5,769,865
-£3,615,450		-£3,635,511		-£3,655,960		-£3,676,525		-£3,697,206		-£3,718,002		-£3,738,916		-£3,759,948		£1,988,768
-£	20,061	-£	20,450	-£	20,565	-£	20,680	-£	20,797	-£	20,914	-£	21,031	-£	21,150	-£	21,269
-£3,635,511		-£3,655,960		-£3,676,525		-£3,697,206		-£3,718,002		-£3,738,916		-£3,759,948		-£3,781,097		£1,967,499

Figure 8.12 Extract from cash flow: letting-up and sales stage

Monthly Totals	=SUM(D13:D42)	=SUM(E13:E42)	=SUM(F13:F42)
Running Total	=D43	=D46+E43	=E46+F43
Interest Charge	0	=(D44)*Interest_Rate_Monthly	=(E44+E45)*Interest_Rate_Monthly
Cumulative Totals (Including interest)	=D44+D45	=E44+E45	=F44+F45

Figure 8.13 Extract from cash flow showing formulas

finance rate for performance analysis purposes but some developers prefer to work with the non-discounted figure which is why I have included both a gross and discounted column in the analysis of results table (Figure 8.8).

The discounted cash flow calculation

The DCF approach uses the same cash flow structure as the residual cash flow but the calculation is different (Figure 8.14). A discount, or present value, factor, based upon the cost of finance, is calculated and used to find the present value of each of the monthly cash flows. The interest charge per period is not calculated.

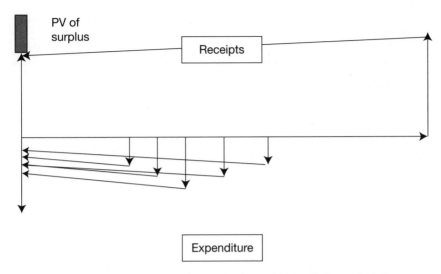

Figure 8.14 The calculation principles involved in a DCF cash flow calculation. Each individual cash flow, both costs and receipts, are discount back to the present, without the calculation of interest. The discount rate used is normally the cost of finance

The calculation for our project is presented in the cash flow extracts in Figures 8.15 to 8.18.

It should be noted that discounted value of the cash flow is identical to the discounted value of the profit in the residual cash flow. The disadvantage of the DCF approach to the developer is that the maximum, real project cash exposure is not revealed, as it is in the residual cash flow. The advantages are that the calculation, if done manually or in a self-created Excel sheet, is easier (which means there is less potential for error), and that it is possible to calculate

Monthly Totals	-£ 745,800	£ -	£ -	-£ 15,000	-£ 12,000	£ -	-£ 52,950
Discount Factor	1.000	0.994	0.989	0.983	0.978	0.972	0.967
Net Value of Cash Flow	-£ 745,800	£ -	£ -	-£ 14,750	-£ 11,734	£ -	-£ 51,198
Cumulative Totals	-£ 745,800	-£ 745,800	-£ 745,800	-£ 760,550	-£ 772,283	-£ 772,283	-£ 823,481

Figure 8.15 Extract from cash flow: land purchase and pre-construction/planning stage

-£ 111,020	-£ 166,530	-£ 277,550	-£ 277,550	-£ 319,305	-£ 271,145	-£ 392,775	-£ 497,325	-£ 309,200
0.961	0.956	0.951	0.945	0.940	0.935	0.930	0.924	0.919
-£ 106,745	-£ 159,222	-£ 263,886	-£ 262,410	-£ 300,199	-£ 253,495	-£ 365,153	-£ 459,764	-£ 284,249
-£ 930,226	-£1,089,449	-£1,353,335	-£1,615,745	-£1,915,944	-£2,169,439	-£2,534,592	-£2,994,356	-£3,278,605

Figure 8.16 Extract from cash flow: construction stage

-£	49,050	£	-	£	-	£	-	£	-	£	-	£	-	£	-	£5,769,865
	0.914		0.909		0.904		0.899		0.894		0.889		0.884		0.879	0.874
-£	44,840	£	-	£	-	£	-	£	-	£	-	£	-	£	-	£5,043,130
-£3,323,445		-£3,323,445		-£3,323,445		-£3,323,445		-£3,323,445		-£3,323,445		-£3,323,445		-£3,323,445	£1,719,686

Figure 8.17 Extract from cash flow: letting-up and sales stage

Monthly Totals	=SUM(D13:D42)	=SUM(E13:E42)	=SUM(F13:F42)
Discount Factor	=(1+Interest_Rate_Monthly)^-DCF!D11	=(1+Interest_Rate_Monthly)^-DCF!E11	=(1+Interest_Rate_Monthly)^
Net Value of Cash Flow	=D44*D43	=E44*E43	=F44*F43
Cumulative Totals	=D45	=D46+E45	=E46+F45

Figure 8.18 Extract from cash flow showing formulas

-£	49,050	£	-	£	-	£	-	£	-	£	-	£	-	£	-	£5,769,865
	0.593		0.574		0.555		0.537		0.520		0.503		0.487		0.471	0.456
-£	29,072	£	-	£	-	£	-	£	-	£	-	£	-	£	-	£2,632,850
-£2,632,850		-£2,632,850		-£2,632,850		-£2,632,850		-£2,632,850		-£2,632,850		-£2,632,850		-£2,632,850	£	-

Finance	
IRR (Annual)	39.88%
IRR (Monthly)	3.32%

Figure 8.19 Project IRR calculation

the Internal Rate of Return (IRR) for the project, an important measure for investors. The IRR is the interest rate that produces an NPV of zero (Figure 8.19).

In this case, the IRR is 39.88 per cent (and has been calculated using the goal seek function of Excel, a useful feature).

Note, given that the IRR is calculated off the gross monthly cash flow figures, an IRR figure can be calculated for the residual cash flow as well.

Conclusions

There has been considerable debate about the merits of the residual and cash flow approaches of both types. It used to be the case that the residual approach was used for initial appraisals, where a quick answer was required and the deficiencies of the approach (the gross simplification of the calculation of interest and the inability to model realistic time-sensitive elements in the developments) could be accepted. Cash flows, on the other hand, although more accurate, were slower to construct and required more detail, and were therefore applied to appraisals done later in a project and as a project management tool.

This debate has largely been overtaken by events. The widespread availability of spreadsheets and proprietary software, all of which use cash flows as the primary calculation tool, has largely made any argument over the merits of each irrelevant.

One area which is relevant to all appraisals is something that I have already touched on, the sensitivity of the outcome of appraisals to certain aspects of the assumptions. This means that all feasibility studies should include some kind of sensitivity analysis.

9 Sensitivity analysis – risk and uncertainty in development projects

Property development is a risky activity. We can tell that from the level of return (i.e. the 'standard' 20 per cent profit margin used in commercial project appraisal) that is assumed. These returns can be obtained, indeed returns in excess of this are quite possible, but, similarly, it is almost equally possible to see heavy losses from a development project.

Why is this the case? Well, to start with, development projects tend to be over relatively long timescales and, therefore, the decision to go ahead and/or the amount to bid for the development site requires judgements to be made about cost and values which are going to occur in the future, frequently indeed 2–5 years in the future. These factors are very hard to predict over these timeframes.

A second problem is that the development appraisal itself is highly sensitive to the key input data, i.e. the assumptions which are themselves uncertain. The reason for this is that the outcome of the development appraisal is actually the marginal difference of the ratio between costs and value. A slight change in this ratio can cause a major percentage difference in this marginal outcome. The key variables are anything to do with value – rents, yields and sale prices. A fairly small reduction in value (say 5 per cent) has a disproportionate effect on land value or development profitability – sometimes 50 per cent or more.

Risk has two aspects. Something can have a low risk of happening (such as being involved in an aircraft crash) but severe consequences (death, serious injury or long-term trauma). Other things can have a high risk of occurrence (e.g. the risk of catching a cold) but insignificant consequences for most (minor discomfort). Property development is subject to factors that have a high risk of variance (rents, yields and sale values are rarely static) but huge stakes and potential consequences because of the high capital sums involved.

Successful developers have to be risk takers. They also have to be optimists – if you are too pessimistic and too cautious the nature of the development appraisal means that you would never have a scheme projected to make a profit nor be able to outbid the market for a piece of land. This does not mean, however, that developers and developer advisers should blindly ignore risk; risk needs to be identified and managed right from the start.

I believe that there are three components to this in the appraisal process.

Firstly, it must be accepted that the outcome of a single development appraisal is just one possible answer. It is up to the person or body constructing it to test and retest the assumptions made for soundness and reasonableness wherever possible. The more sound the underlying assumptions are, the more reliable the outcome will be.

Secondly, the tools used to carry out the development appraisal should have mechanisms to test the sensitivity of the appraisal to variations in the input variable.

Finally, the person carrying out the feasibility study should have the skills and knowledge to use this sensitivity analysis properly. All too often there is a temptation to simply carry out a 'standard' sensitivity analysis – varying the rent and yield by 10 per cent and recording the results. I would argue that this is just going through the motions, it is not truly 'analysis'. The skill in sensitivity analysis is knowing what questions to ask to ensure that the risk in the development (and to the developer) is fully appreciated and explored.

The first of these components is down to the developer/appraiser in an individual project. They must understand the market, understand the forces that act on it, understand the product they are intending to produce, and this can only really come from both experience and good research.

The latter two components are different because they are concerned with the tools and methodology, i.e. how the tools are used, and this is the area we can examine. Fortunately, the main software products used by practitioners in the market place, Argus Developer and Estate Master DF have a good sensitivity analysis module, and later sections will look at how to get the best out of it, but for the remainder of this section I will be using Excel to explore the characteristics of sensitivity analysis.

Sensitivity analysis can be carried out in a number of different ways with each method having different utilities and uses. Each form of sensitivity also a number of sub types. These are:

- Simple sensitivity

  - Single variable analysis – changing variables one-by-one by fixed similar amounts. This is done to determine which variables the developer has to pay the most attention to.
  - Single variable analysis for the break-even point – in this analysis the object is to discover the value of the variable that will reduce the profitability of the scheme to zero. Although this is not that far removed from the basic simple sensitivity approach it is sometimes more informative for the developer.

- Scenario-based sensitivity analysis

  - Basic scenarios – determining which groups of variables might move together and exploring the effect that this has on the development profitability or land value compared with the base assumptions.

- Probability-linked scenario analysis – basically an extension of the above but with an attempt made to ascribe the probability of the scenarios occurring. Although this in itself can be subjective it can really inform the development decision process.

- Simulation

  - This is running a Monte Carlo type analysis on the development variables. This is an advanced technique, not currently available in any of the proprietary models (the closest being in Estate Master DF, where there is a probability based function) and only available on Excel via add-ins or by self-created programmes.

This book will look at the use of simple sensitivity and scenarios. For all examples we will use just the input assumptions and results panels, using as the base appraisal the cash flow example from the preceding section. The base result figures are shown in Figure 9.1.

As noted in the text above, simple sensitivity testing is achieved by changing one variable at a time and noting the results of the change. Normally this is done by changing each variable by a fixed value or percentage figure. In this case the rent (Figures 9.2 and 9.3) and the yield (Figures 9.4 and 9.5) have been changed by the same 10 per cent figure.

In the case of rental values the 10 per cent reduction has decreased net profits by 29.32 per cent whilst the 10 per cent reduction in yields has seen profits fall by about 28 per cent (an increase in yield always reduces values as the income multiplier falls as the required yield rises). This level of sensitivity is quite normal in development projects; indeed often the degree of sensitivity is greater than this.

This process can (and usually is) repeated for each key variable in the project though the results tend to be the same. Anything that affects value tends to have the greatest impact – rent, yields, rent frees, time-to-let or sell, etc. – though construction costs, land price, project duration and interest rates can also be significant. The overall process informs the developer/appraiser where

	Gross	Discounted
Profit	£1,967,499	£1,719,686
Return on Cost	47.06%	41.14%
Return on Development Value	31.87%	27.86%
Development Yield	10.00%	10.00%
IRR (monthly)	3.323%	3.323%
IRR (annual)	39.88%	39.88%
Rent Cover	56.47 Months	49.36 Months

Figure 9.1 Base assumption results

Development Assumptions

Finance	
Interest Rate (Annual)	6.75%
Interest Rate (Monthly)	0.56%
Arrangement Fee	£25,000

Timescales	
Planning/preconstruction	6 Months
Construction	9 Months
Letting/Sale	6 Months

Land	
Acquisition Cost	£680,000.00

Areas	Gross	Net	Construction Costs	Rental Values	Yields	Rent Free
Office	1,300 m²	1,079 m²	£800	£167	6.50%	6 Months
Retail Warehouse	1,000 m²	950 m²	£600	£207	6.25%	6 Months
Landscaping (hard & soft)	3,000 m²	3,000 m²	£250			

Fees		Contingency	5%		
Architect	4.00%	Landscape Architect	2.00%	Land Legal	1.00%
Quantity Surveyor	2.00%	Planning Consultant	£12,000	Letting Agent	10.00%
Structural Engineer	1.00%	Planning Fee	£15,000	Letting Legal	5.00%
M&E Engineer	1.00%	Sale Agent	1.00%	Stamp Duty	4.00%
				Incoming purchasers Costs	6.00%
Project Manager	1.00%	Sale Legal	1.00%		
Site Safety	0.75%	Land Agent	0.75%	Marketing	£25,000

Figure 9.2 Rental values reduced by 10%

Profit	Gross	Discounted
	£1,390,513	£1,215,373
Return on Cost	33.59%	29.36%
Return on Development Value	25.03%	21.88%
Development Yield	9.09%	9.09%
IRR (monthly)	2.632%	2.632%
IRR (annual)	31.59%	31.59%
Rent Cover	44.34 Months	38.76 Months

Figure 9.3 Results of rental values reduced by 10%

Development Assumptions

Finance	
Interest Rate (Annual)	6.75%
Interest Rate (Monthly)	0.56%
Arrangement Fee	£25,000

Timescales	
Planning/preconstruction	6 Months
Construction	9 Months
Letting/Sale	6 Months

Land	
Acquisition Cost	£680,000.00

Figure 9.4 Capitalisation yields increased by 10%

Areas	Gross	Net	Construction Costs	Rental Values	Yields	Rent Free
Office	1,300 m²	1,079 m²	£800	£185	7.15%	6 Months
Retail Warehouse	1,000 m²	950 m²	£600	£230	6.88%	6 Months
Landscaping (hard & soft)	3,000 m²	3,000 m²	£250			

Fees					
Contingency	5%				
Architect	4.00%	Landscape Architect	2.00%	Land Legal	1.00%
Quantity Surveyor	2.00%	Planning Consultant	£12,000	Letting Agent	10.00%
Structural Engineer	1.00%	Planning Fee	£15,000	Letting Legal	5.00%
M&E Engineer	1.00%	Sale Agent	1.00%	Stamp Duty	4.00%
Project Manager	1.00%	Sale Legal	1.00%	Incoming purchasers Costs	6.00%
Site Safety	0.75%	Land Agent	1.00%	Marketing	£25,000

	Gross	Discounted
Profit	£ 1,416,134	£ 1,237,767
Return on Cost	33.97%	29.69%
Return on Development Value	25.24%	22.07%
Development Yield	10.03%	10.03%
IRR (monthly)	2.665%	2.665%
IRR (annual)	31.98%	31.98%
Rent Cover	40.64 Months	35.52 Months

Figure 9.5 Results of capitalisation yields increased by 10%

they are most vulnerable and where they need to concentrate their efforts either on improving the accuracy of the value estimate or in reducing the risk in the actual project.

The actual value of simple sensitivity, is, however limited as it is rare that individual values move on their own. Even exploring what values are required to extinguish profitability are of limited utility (Figures 9.6 and 9.7). The single figure of £83 per m² is almost unthinkable. This exercise can, however, take the exploration of risk for the developer further.

What is much more useful is to take a scenario-based approach, because this is much more realistic. Input values will move together where market conditions improve or deteriorate. It can be seen that small, linked movements can have a very significant effect. This is explored in Figures 9.8 and 9.9 where the effects of what might happen in a downturn are modelled. If markets turn down the impact tends to be on both rents *and* yields. In addition incentives to tenants such as rent free periods increase and interest rates rise as bank lending becomes riskier.

The net effect of changing these four variables is that profits are reduced by a dramatic 67 per cent.

This approach to sensitivity, creating scenarios based on 'what-if' questions and reasoned thought is far more effective and informative to the developer/ appraiser. The Excel user is helped by the existence of a scenario manager module in the programme (which also has a 'goal-seek' function that has already been used in the calculation of the profit extinguishment) which enables the exploration of changing multiple variables and the creation of multiple scenarios – indeed an almost infinite number if required (see Figures 9.10 to 9.12).

In Part Two of this book, the issues with using Excel models are discussed further; however there is no doubt that this ability to explore risk and scenarios is a huge advantage, and the flexibility of exploration that the system offers is greater than that with the proprietary models. Scenarios themselves offer many advantages over simple sensitivity to the developer. They can also be extended by applying probability values to the individual scenarios (see Figure 9.13). Here the developer/appraiser has suggested that there is a 1 in 2 chance (i.e. 50 per cent) of the base (normal) scenario occurring whilst the high (market improvement) and low (market deterioration) each have a 1 in 4 chance of happening. Due to the greater impact of the downside values, this produces a lower weighted average predicted profit.

Whilst these probability-based scenarios are only as good as the quality of the input assumptions used (the probability values themselves tend to be subjective), the whole approach does inform the development feasibility process greatly, and scenarios seem to be the minimum level of sensitivity analysis that should be done in a development feasibility study.

Development Assumptions

Finance	
Interest Rate (Annual)	6.75%
Interest Rate (Monthly)	0.56%
Arrangement Fee	£25,000

Timescales	
Planning/preconstruction	6 Months
Construction	9 Months
Letting/Sale	6 Months

Land	
Acquisition Cost	£680,000.00

Areas	Gross	Net	Construction Costs	Rental Values	Yields	Rent Free
Office	1,300 m²	1,079 m²	£800	£185	6.50%	6 Months
Retail Warehouse	1,000 m²	950 m²	£600	£83	6.25%	6 Months
Landscaping (hard & soft)	3,000 m²	3,000 m²	£250			

Fees		Contingency	5%		
Architect	4.00%	Landscape Architect	2.00%	Land Legal	1.00%
Quantity Surveyor	2.00%	Planning Consultant	£12,000	Letting Agent	10.00%
Structural Engineer	1.00%	Planning Fee	£15,000	Letting Legal	5.00%
M&E Engineer	1.00%	Sale Agent	1.00%	Stamp Duty	4.00%
				Incoming purchasers Costs	6.00%
Project Manager	1.00%	Sale Legal	1.00%	Marketing	£25,000
Site Safety	0.75%	Land Agent	1.00%		

Figure 9.6 Required reduction in retail warehouse rents to extinguish development profitability

Profit	Gross		Discounted	
	£	0	£	0
Return on Cost	0.00%		0.00%	
Return on Development Value	0.00%		0.00%	
Development Yield	6.88%		6.88%	
IRR (monthly)	0.562%		0.562%	
IRR (annual)	6.75%		6.75%	
Rent Cover	0.00 Months		0.00 Months	

Figure 9.7 Development profitability extinguished

Development Assumptions

Finance		
Interest Rate (Annual)	8.00%	
Interest Rate (Monthly)	0.67%	
Arrangement Fee	£25,000	

Areas	Gross	Net	Construction Costs	Rental Values	Yields	Rent Free
Office	1,300 m²	1,079 m²	£800	£167	7.15%	12 Months
Retail Warehouse	1,000 m²	950 m²	£600	£207	6.88%	12 Months
Landscaping (hard & soft)	3,000 m²	3,000 m²	£250			

Contingency 5%

Fees					
Architect	4.00%	Landscape Architect	2.00%	Land Legal	1.00%
Quantity Surveyor	2.00%	Planning Consultant	£12,000	Letting Agent	10.00%
Structural Engineer	1.00%	Planning Fee	£15,000	Letting Legal	5.00%
M&E Engineer	1.00%	Sale Agent	1.00%	Stamp Duty	4.00%
				Incoming purchasers Costs	6.00%
Project Manager	1.00%	Sale Legal	1.00%	Marketing	£25,000
Site Safety	0.75%	Land Agent	1.00%		

Timescales	
Planning/preconstruction	6 Months
Construction	9 Months
Letting/Sale	6 Months

Land	
Acquisition Cost	£680,000.00

Figure 9.8 Inputs for a scenario modelling a market downturn

Profit	Gross	Discounted
Profit	£652,904	£556,664
Return on Cost	14.91%	12.72%
Return on Development Value	12.91%	11.01%
Development Yield	8.61%	8.61%
IRR (monthly)	1.708%	1.708%
IRR (annual)	20.50%	20.50%
Rent Cover	20.79 Months	17.73 Months

Figure 9.9 Results of scenario

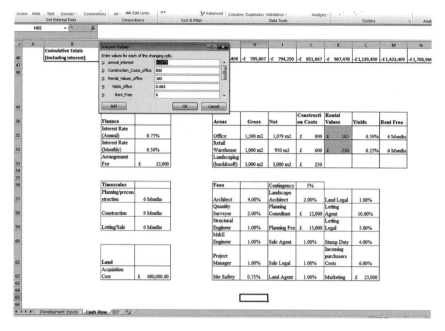

Figure 9.10 Using Excel's scenario manager to define variables

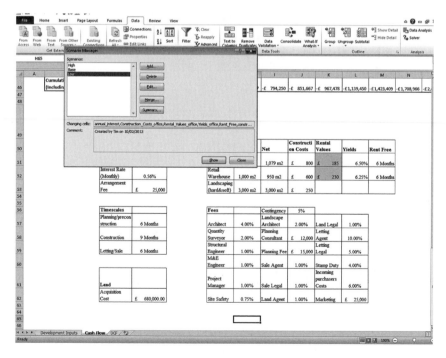

Figure 9.11 Using Excel's scenario manager to model multiple development scenarios

	A	B	C	D	E	F	G	H
1								
2		**Scenario Summary**						
3				Current Values:	High	Base	Low	
5		Changing Cells:						
6			annual_interest	7.25%	6.25%	6.75%	7.25%	
7			Construction_Costs_office	£ 850	£ 790	£ 800	£ 850	
8			Rental_Values_office	£ 170	£ 200	£ 185	£ 170	
9			Yields_office	7.00%	6.00%	6.50%	7.00%	
10			Rent_Free	9 Months	3 Months	6 Months	9 Months	
11			construction_costs_retail	£ 625	£ 590	£ 600	£ 625	
12			rental_values_retail	£ 200	£ 250	£ 230	£ 200	
13			Yields_retail	6.75%	6.00%	6.25%	6.75%	
14			rent_free_retail	12 Months	3 Months	6 Months	12 Months	
15		Result Cells:						
16			PV Profit	£ 587,342	£ 2,664,865	£ 1,719,686	£ 587,342	
17			Profit on Cost	13.33%	65.67%	41.14%	13.33%	
18			IRR	20.01%	52.87%	39.88%	20.01%	
19		Notes: Current Values column represents values of changing cells at						
20		time Scenario Summary Report was created. Changing cells for each						
21		scenario are highlighted in gray.						
22								

Figure 9.12 Using Excel's scenario manager to model multiple development scenarios: results table

	High	Base	Low
Scenario Summary	Created by Tim on 10/02/2013	Created by Tim on 10/02/2013	Created by Tim on 10/02/2013
Changing Cells:			
annual_interest	6.25%	6.75%	7.25%
Construction_Costs_office	£ 790	£ 800	£ 850
Rental_Values_office	£ 200	£ 185	£ 170
Yields_office	6.00%	6.50%	7.00%
Rent_Free	3 Months	6 Months	9 Months
construction_costs_retail	£ 590	£ 600	£ 625
rental_values_retail	£ 250	£ 230	£ 200
Yields_retail	6.00%	6.25%	6.75%
rent_free_retail	3 Months	6 Months	12 Months
Result Cells:			
PV Profit	£ 2,664,865	£ 1,719,686	£ 587,342
Profit on Cost	65.67%	41.14%	13.33%
IRR	52.87%	39.88%	20.01%
Probability Of Occurance	25%	50%	25%
Weighted Result	£ 666,216	£ 859,843	£ 146,835
		Weighed Average Profit £	1,672,895

Figure 9.13 Probability extension to scenario modelling

10 Conclusions to Part One

Part One started off with a definition of development. This definition, and the later sections that gave an overview of the market, illustrates what a broad topic development is and how many different players there are involved in it.

This creates difficulty for a book that is focussed on one aspect – appraisal. It is inevitable that in the early sections, when we are looking at the overall principles involved in the process, that this has to be done in rather general terms. I hope that the preceding sections have captured the essence of the appraisal process – starting from establishing what goals the developer and development has, through determining the constraints and parameters of the development and then looking at what areas the appraiser/developer needs to collect information on. What this finally leads to is how this information is applied, the structure and mechanics of the calculations – be it residual or cash flow structure – and the sensitive, unstable nature of the calculation outcome.

In Part Two, we will move from the general to the more specific, looking at how appraisal is now done in practice using the tools available to the modern appraiser/developer. This will involve more specific case studies but the reader should see the same process and issues that have been explored in Part One carried through into the practice section.

Part Two

Development appraisal in practice

In Part Two, the principles of development appraisal covered in Part One are applied using the tools available to the appraiser/developer. Part Two will, therefore, concentrate on the mechanics of constructing appraisals. In doing this, however, it is assumed that the in-depth research on the myriad aspects that make up even a simple development project will already have been done.

This part is split into three chapters, concentrating on the three most commonly used tools in the industry – Microsoft Excel, Argus Developer and Estate Master DF. There are other tools available, other types of spreadsheet and other proprietary software models include Caldes, Prodeveloper and Kel systems; however the vast majority of appraisals will be undertaken using these three systems.

The three chapters are slightly different. Those for Argus and Estate Master include an introductory outline to the systems followed by case studies, though some critical analysis of the software is made. The Excel section uses some examples but concentrates more on the risks involved in self-created calculations.

11 Development appraisal using MS Excel

We have already covered Excel models in both the chapter on the basic mechanics of appraisal (both the residual model illustrated and the two forms of cash flow were calculated using Excel) and in the chapter on sensitivity analysis. This illustrates the main characteristics of the programme: its flexibility and utility – it is an immensely powerful tool that can be applied to virtually any situation.

I used to be a strong proponent of Excel spreadsheets, frequently using and creating them, employing them widely in both practice and in the classroom. I am now more wary of them, even though they are still probably the most widely used tool in the industry. In early 2012 I found myself embroiled in a debate on an internet real estate discussion site about the merits of proprietary software versus Excel.

I came across the site accidentally whilst checking a reference. I was interested in the debate that had clearly been going backwards and forwards for some weeks and, always enjoying an argument, decided I would join in. Having been heavily involved in this area for a dozen years or so and having originally been a great advocate of Excel, I found it odd to be on the other side of the fence arguing with the site owner. He did make some excellent points some of which I quote here:

> I have a good friend who is an engineer for a leading aerospace-component design and manufacturing facility here in southern California. They produce nacelles mostly, but that's beside the point. Their parts are used by both Boeing and Airbus. Their work requires modelling the physics acting on their parts and extensive and sophisticated testing is required by regulators and for simply quality control purposes. What do they use? Microsoft Excel with Visual Basic. The planes you and I ride in were in part designed and tested with Excel.
>
> I have a colleague who works on Wall Street creating and trading financial derivatives for a leading investment bank. His whole firm (and industry) use Excel. Why? Because of its power and breadth and because they – like every other MBA from a top university – used Excel for their finance and statistics classes. All the leading texts (particularly Bodie/

Kane/Marcus) in finance go through examples in Excel. And this isn't simple discounting we're talking about but CAPM, linear regression, options pricing, etc. Excel is the standard.

As much as I like to think that real estate analysis and valuation demands brilliance and a firm grasp of complicated and arcane mathematics/statistics (and good looks to boot!), it simply doesn't. Beyond logic (if this, then this … etc) and high school math (but not even including calculus), commercial real estate analysis requires only an understanding of Present Value (mathematically speaking). The modelling we do for real estate is child's play compared to the capabilities of Excel, as evidenced by its much more rigorous applications.

(Landon M. Scott http://incomepropertyanalytics.com/
alternative-to-argus/#comment-896 Accessed
7 May 2012, printed with permission)

It is natural, then, that Excel be applied to property development. It looks like the ideal tool for the job and, indeed, many practitioners pride themselves on their Excel abilities and the sophistication of the models they use.

There are, however, issues with using Excel that the practitioner should be aware of, and this chapter concentrates on these.

For reasons of accuracy there has been a general pressure in the development industry to move from the simple residual to cash flow approaches. However, there are problems with cash flow feasibility studies which have meant that some sections of the industry have resisted their use. The first principal complaint about the models is that they are time consuming to produce from scratch, particularly compared with residual models. Once a template model has been developed however, this time can be reduced. Nonetheless, there are inherent risks of error when existing models are adapted to meet the needs of different feasibility studies. The second major complaint is the level of detail required in the assumptions that go into the construction of the cash flow template. Again there are elements of truth in this; the models are far more complex and transparent yet many of the assumptions required can reasonably be made from past experience from similar projects and are not that far removed from the sweeping, broad-brush ones made in the residual models.

The final major complaint is one that is valid; the increased complexity of the cash flow models means that there is more risk of error creeping in. These are not so much errors of assumption but more simple mistakes in calculation or cell reference. Spreadsheet cash flow models have to be very carefully audited and, often, there is insufficient time to do this. This is undeniable and is why many are advocates of using proprietary models where these types of errors can be virtually eliminated.

There is, however, more than one type of spreadsheet model used in development appraisal practice, mainly using the ubiquitous Microsoft Excel, although there is some use of Open Office and Libre Office applications. Essentially, experience has shown that the models fall into four broad types:

1 an automated, 'Active' residual model reproduced in Excel
2 'simple' self-created calculation sheets lacking time specific dialogue references
3 more sophisticated self-created sheets with time specific dialogue references
4 professional created complex sheets.

These models' usability and propensity to be vulnerable to errors vary and it is therefore important to look at the outline characteristics of each.

The first type was illustrated in the chapter on the residual model. They are spreadsheets, albeit simple ones, in that the calculation is done behind the cell where the results are displayed (Figure 11.1).

As these reproduce much of the weaknesses of the residual models, there seems little attraction in using them. Quite often, however, the residual layout is used to display the results of the calculation that occurs in a cash flow somewhere else. The residual layout has distinct advantages in presenting results; indeed most proprietary systems such as Argus Developer and Estate Master DF use this approach.

The second type is typical of the majority of self-created sheets, where the user, probably self-taught on Excel, has produced an often unique or tailored spreadsheet to conduct a particular project. This is normally done in essentially a two-stage process. In the first stage we need to establish the calculation assumptions – the broad timescales, the net and built areas, the likely rent and sale values, the likely interest rates etc. This was what was done in the examples illustrating the two cash flow approaches in Part One, and they are reproduced again here for completeness (Figures 11.2 to 11.6).

I believe that most people involved in the industry will have constructed similar spreadsheets. These 'one-off', specifically constructed models may have the sophistication of calculation by reference to named cells but normally we see these calculations manually distributed into a static time framework (see extract in Figure 11.7) rather than using the more complex type of dynamic timing as covered in the models below which are more time consuming to create and require more in-depth auditing for errors.

◢	A	B	C	D	E	F
1						
2	Preconstruction	6 Months				
3	Construction	9 Months				
4	Letting	9 Months				
5	Discount Rate	10%				
6	Value of Scheme					
7			m^2	£		
8	Retail Net Area		212.5	£	200	=D8*C8	
9	Office Net Area		830	£	175	=D9*C9	
10	Total Income				=SUM(E8:E9)	
11	Yield			7%	=(1/D11)	=E11*E10
12	Capital Value					
13						

Figure 11.1 Calculation of value on completion

Development Assumptions

Finance	
Interest Rate (Annual)	6.75%
Interest Rate (Monthly)	0.56%
Arrangement Fee	£25,000

Areas	Gross	Net	Construction Costs	Rental Values	Yields	Rent Free
Office	1,300 m²	1,079 m²	£800	£185	6.50%	6 Months
Retail Warehouse	1,000 m²	950 m²	£600	£230	6.25%	6 Months
Landscaping (hard & soft)	3,000 m²	3,000 m²	£250			

Timescales	
Planning/preconstruction	6 Months
Construction	9 Months
Letting/Sale	6 Months

Fees					
Contingency	5%				
Architect	4.00%	Landscape Architect	2.00%	Land Legal	1.00%
Quantity Surveyor	2.00%	Planning Consultant	£12,000	Letting Agent	10.00%
Structural Engineer	1.00%	Planning Fee	£15,000	Letting Legal	5.00%
M&E Engineer	1.00%	Sale Agent	1.00%	Stamp Duty	4.00%
				Incoming purchasers Costs	6.00%
Project Manager	1.00%	Sale Legal	1.00%		
Site Safety	0.75%	Land Agent	0.75%	Marketing	£25,000

Land	
Acquisition Cost	£680,000.00

Figure 11.2 Development assumptions used in calculation

Months	0	1	2	3	4	5	6
Costs/Values							
Rental Income - Office							
Rent Free - Office							
Sale - Office							
Rental Income - Retail							
Rent Free - Retail							
Sale - Retail							
Purchasers Costs							
Land acquisition	-£ 680,000						
Stamp duty	-£ 27,200						
Construction - Office							
Construction - Retail							
Construction - Landscape							
Contingency							
Architect							-£ 32,800
Quantity Surveyor							-£ 11,950
Structural Engineer							-£ 4,100
M&E Engineer							-£ 4,100
Project Manager							
Site Safety							
Landscape Architect							
Planning Consultant					-£ 12,000		
Planning Fee				-£ 15,000			
Sale Agent							
Sale Legal							
Land Agent	-£ 6,800						
Land Legal	-£ 6,800						
Letting Agent							
Letting Legal							
Loan arrangement fee	-£ 25,000						
Marketing							
Monthly Totals	-£ 745,800	£ -	£ -	-£ 15,000	-£ 12,000	£ -	-£ 52,950
Running Total	-£ 745,800	-£ 745,800	-£ 749,995	-£ 769,214	-£ 785,456	-£ 789,807	-£ 847,200
Interest Charge	£ -	-£ 4,195	-£ 4,219	-£ 4,242	-£ 4,351	-£ 4,443	-£ 4,468
Cumulative Totals (Including interest)	-£ 745,800	-£ 749,995	-£ 754,214	-£ 773,456	-£ 789,807	-£ 794,250	-£ 851,667

Figure 11.3 Pre-construction phase of cash flow

The third type of spreadsheet is better because of the introduction of logical statements that control time assumptions in the appraisal. An example of this was presented by Peter Brown in a series of articles for the Institute of Revenues Rating and Valuation (IRRV) in 2011 (http://www.irrv.net/home/item.asp?id=1319 where the models displayed in Figures 11.8 and 11.9 can be downloaded).

These type two spreadsheets have a much more sophisticated treatment of time. The development time framework shrinks or contracts according to the entry made into cell B17 (see Figure 11.9). Other, time-specific, items are entered and distributed into the cash flow using the formulas entered into the start–end boxes. The strengths of these types of spreadsheet are their ease and flexibility of use once created and the lack of need to re-audit when timing assumptions are changed. Their weaknesses are twofold; firstly they need more time and underlying knowledge to be created and audited for errors initially. Secondly, they are vulnerable to later changes; their complexity can

7	8	9	10	11	12	13	14	15
-£ 104,000	-£ 156,000	-£ 260,000	-£ 260,000	-£ 156,000	-£ 104,000			
				-£ 90,000	-£ 150,000	-£ 180,000	-£ 90,000	-£ 90,000
						-£ 187,500	-£ 375,000	-£ 187,500
-£ 5,200	-£ 7,800	-£ 13,000	-£ 13,000	-£ 12,300	-£ 12,700	-£ 18,375	-£ 23,250	-£ 13,875
				-£ 16,400				
				-£ 23,900				
				-£ 8,200				
				-£ 8,200				
-£ 1,040	-£ 1,560	-£ 2,600	-£ 2,600	-£ 2,460	-£ 2,540	-£ 1,800	-£ 900	-£ 900
-£ 780	-£ 1,170	-£ 1,950	-£ 1,950	-£ 1,845	-£ 1,905	-£ 1,350	-£ 675	-£ 675
						-£ 3,750	-£ 7,500	-£ 3,750
								-£ 12,500
-£ 111,020	-£ 166,530	-£ 277,550	-£ 277,550	-£ 319,305	-£ 271,145	-£ 392,775	-£ 497,325	-£ 309,200
-£ 962,687	-£1,134,008	-£1,417,000	-£1,700,959	-£2,028,271	-£2,309,029	-£2,713,267	-£3,223,645	-£3,548,180
-£ 4,791	-£ 5,442	-£ 6,409	-£ 8,007	-£ 9,613	-£ 11,463	-£ 13,053	-£ 15,336	-£ 18,219
-£ 967,478	-£1,139,450	-£1,423,409	-£1,708,966	-£2,037,884	-£2,320,492	-£2,726,320	-£3,238,980	-£3,566,400

Figure 11.4 Construction phase of cash flow

hide errors that have come about by a hurried change of the fundamental formula, something that can frequently happen on a project to deal with a specific change in assumptions. Although formulas can be write-protected, often the security process can be bypassed or turned off completely.

The fourth type of spreadsheet is the more sophisticated, often professionally produced models. An example is the UK Government's Homes and Communities Agency (HCA) Development Appraisal Tool (DAT) (Figure 11.10) (downloadable from http://www.homesandcommunities.co.uk/ourwork/development-appraisal-tool).

This tool is essentially a highly protected Excel spreadsheet which allows the user access only to the input screens. The main reason that the HCA use this tool (and require developers working with them and their local housing authority partners and social housing providers to use it as well) is that it addresses one of the other key issues with self-created spreadsheet models: consistency. The HCA administer grants and the development and acquisition of social housing that are provided by the private sector and operated by Housing Associations, non-profit making organisations working closely with the local

16	17	18	19	20	21	22	23	24
								-£ 99,808
								£3,071,000
								-£ 109,250
								£3,496,000
								-£ 394,020
-£ 16,400								
-£ 11,950								
-£ 4,100								
-£ 4,100								
								-£ 65,670
								-£ 65,670
								-£ 41,812
								-£ 20,906
-£ 12,500								
-£ 49,050	£ -	£ -	£ -	£ -	£ -	£ -	£ -	£5,769,865
-£3,615,450	-£3,635,511	-£3,655,960	-£3,676,525	-£3,697,206	-£3,718,002	-£3,738,916	-£3,759,948	£1,988,768
-£ 20,061	-£ 20,450	-£ 20,565	-£ 20,680	-£ 20,797	-£ 20,914	-£ 21,031	-£ 21,150	-£ 21,269
-£3,635,511	-£3,655,960	-£3,676,525	-£3,697,206	-£3,718,002	-£3,738,916	-£3,759,948	-£3,781,097	£1,967,499

Figure 11.5 Post–construction phase of project

	Gross	Discounted
Profit	£1,967,499	£1,719,686
Return on Cost	47.06%	41.14%
Return on Development Value	31.87%	27.86%
Development Yield	10.00%	10.00%
IRR (monthly)	3.323%	3.323%
IRR (annual)	39.88%	39.88%
Rent Cover	56.47 Months	49.36 Months

Figure 11.6 Performance analysis

	B	D	E
36	Sale Legal		
37	Land Agent	=Land_Agent*D20	
38	Land Legal	=Land_Legal*D20	
39	Letting Agent		
40	Letting Legal		
41	Loan arrangement fee	=-Arrangement_Fee	
42	Marketing		
43	Monthly Totals	=SUM(D13:D42)	=SUM(E13:E42)
44	Cumulative Totals (excluding interest)	=D43	=D44+E43
45	Interest Charge	0	=(D44)*Interest_Rate__Monthly
46	Cumulative Totals (Including interest)	=D44+D45	=E44+E45

Figure 11.7 Extract from the type one spreadsheet model showing formulas

	A	B	E	F	H	I
1						
2						
3	**Project Details**					
4						
5	Site Area (ha)	1.00				
6	Gross External Area (sq.m)	10,000				
7	Net Lettable Area (sq.m)	9,000				
8	Development Type	Office				
9						
10	**Development Values**					
11	Rental Value (sq.m)	£100				
12	Rental Value p.a.				£900,000	
13	Yield %	10.00%				
14	Gross Development Value				£9,000,000	
15						
16	**Development Time Frame**					
17	Development period (mths)	24				
18	Start Date	January 2011				
19			Start	End	£	
20	**Construction Costs**					
21	Building Costs (sq.m)	£400	Feb-11	Dec-12	£4,000,000	
22	Landscaping	£20,000	Sep-12	Dec-12	£20,000	
23	Other external costs	£30,000	Jul-12	Dec-12	£30,000	
24	Site Clearance	£20,000	Jan-11	Feb-11	£20,000	
25	Site Preparation	£40,000	Jan-11	Mar-11	£40,000	
26	**Total Construction Costs**				**£4,110,000**	
27						
28	**Fees**					
29	Architects %	10.00%	Mar-11	May-11	£400,000	% of building cost
30	Quantity Surveyors %	5.00%	Apr-11	Jul-11	£200,000	% of building cost
31	Legal Fees on sale %	2.00%	Dec-12	Dec-12	£180,000	% of Gross Development Value
32	Legal Fees on letting %	2.00%	Dec-12	Dec-12	£18,000	% of rent
33	Letting Agents %	2.00%	Dec-12	Dec-12	£18,000	% of rent
34	Civil Engineers %	5.00%	May-11	Jul-11	£200,000	% of building cost
35						
36	Planning Permission	£10,000	Jan-11	Mar-11	£10,000	
37	Building Regulations	£5,000	Jan-11	May-11	£5,000	
38	**Total Fees**				**£1,031,000**	
39						
40	**Other Costs**					
41	Contingencies %	10.00%	Dec-12	Dec-12	£514,100	% of Construction costs and fees
42	**Total Contingencies**				**£514,100**	
43						
44	**Developer's Profit**					
45	Profit %	15.00%			£1,350,000	
46	**Total Developer's Profit**				**£1,350,000**	
47						
48	**Finance**					
49	Finance Rate p.a.	10.00%				
50	Finance Rate (mth)				0.797%	

Read Me | **Inputs** / Residual - Land Value / Residual - Profit / DCF - Land Value / DCF - Prof

Select destination and press ENTER or choose Paste

Figure 11.8 Extract from Peter Brown's Development Appraisal Model

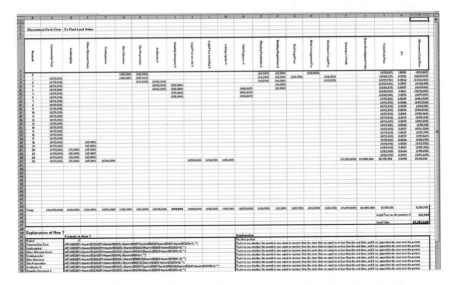

Figure 11.9 Spreadsheet produced by the Peter Brown model including example of formulas and logical time references

Figure 11.10 The initial screen of the HCA's DAT

housing authorities throughout the country. Many of these schemes in deprived regions and cities require assistance and gap funding, the assessment of need for which is on the appraisal. To ensure that the decision making is consistent they needed this universal tool. The HCA website says:

> The HCA's Development Appraisal Tool (DAT) is designed to appraise in detail the viability of an individual site. It takes into account local assumptions for costs and value, and records the dates at which these assumptions impact on a project cashflow over the life cycle of the development. This will help to identify the residual land value or funding deficit. It is intended to be transparent and easy to use by both delivery teams and client organisations.

Using the tool

The HCA is using the tool as part of the Delivery Partner Panel 2 initial procurement to make it easier to compare and benchmark aspects of the bids, and will continue to use the method as part of on-going disposals through the panel in future. The ability to compare bids on the same basis will help to ensure compliance in use of the panel, greater transparency and the use of a single model will generate significant efficiencies in use for both the public and private sectors.

The model can also assist local authority planning teams manage individual site viability negotiations during the planning process to agree an affordable housing mix alongside the impact of Community Infrastructure Levy (CIL) and other planning obligations. The DAT complements the HCA Area Wide Viability model which can be used by local planning authorities to viability test their planning policy – either at local plan stage, or affordable housing policy setting or to establish a CIL charging structure on typical site typologies throughout their area in accordance with the requirements of the National Planning Policy Framework.

Under national planning policy guidance (PPS3) local authorities are expected to maximise affordable housing delivery through the use of developer contributions. The 2011–15 Affordable Homes Programme framework states the HCA's expectation that affordable homes delivered on developer led mixed tenure sites through s106 agreements will be delivered on a nil-grant basis. Where HCA funding is sought on such sites, the DAT model will be used to test the additional homes that grant might deliver. The tool has capacity to test Affordable Rent tenure as well as Social Rent and Shared Ownership tenures alongside Open Market Sale and Private Rented units. The model includes analytical tools including capacity to run sensitivity analysis scenarios.

(HCA, www.homesandcommunities.co.uk/ourwork/
development-appraisal-tool, accessed January 2013)

The DAT has two templates for inputting data, a simple and a complex one, the latter for advanced users, and can model both purely residential schemes and mixed use projects. Some of the input and results screens are illustrated in Figures 11.11 to 11.15.

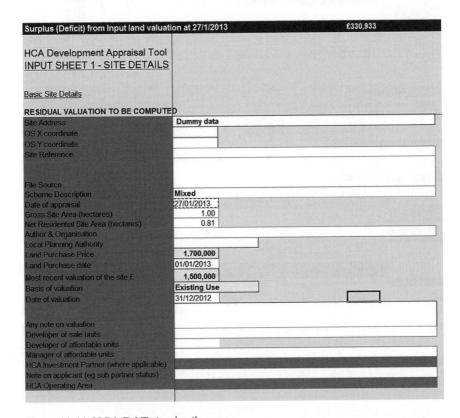

Figure 11.11 HCA DAT site details screen

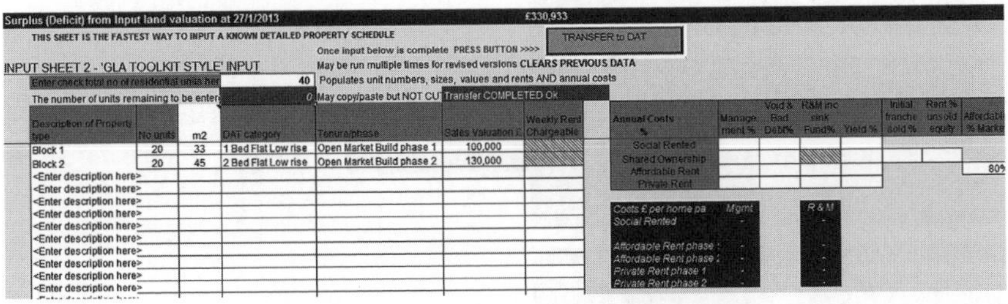

Figure 11.12 Residential property input screen

Surplus (Deficit) from Input land valuation at 27/1/2013　　　£330,933

HCA Development Appraisal Tool
INPUT SHEET 2 - PHASING

Note: A Tenure/Phase must have units entered in order to display on this sheet

	Construction Start Date	Construction End Date	Construction Start Month no.	Construction End Month no.	No. of units in tenure	
OM 1 Build phase 1	01-Mar-13	01-Dec-13	1	10	20	
OM 2 Build phase 2	01-Jun-13	01-Mar-14	4	13	20	
	RP Purchase (transfer) start date	Purchase end date	Start Month	End Month	No. of units in tenure	
	Open Market Sale Start Date	Sale End Date	Start Month	End Month	No. of units in tenure	Monthly Sales rate
OM Sales 1 Build phase 1	01-Oct-13	01-Mar-14	8	13	20	3.33
OM Sales 2 Build phase 2	01-Dec-13	01-Jun-14	10	16	20	2.86

Figure 11.13 HCA DAT time framework entry

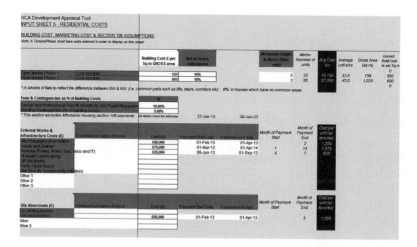

Figure 11.14 HCA DAT cost data entry

Surplus (Deficit) from Input land valuation at 27/1/2013　　　£330,933

HCA Development Appraisal Tool INPUT SHEET 6 - NON-RESIDENTIAL	Dates must be between and	27-Jan-13 27-Jan-23

ASSUMPTIONS by user defined type
Office
Comments here
Size of Office scheme (gross sq m)
Size of Office scheme (net lettable sq m)

Values
Rent (£ psm)
Investor's Yield (%)
Costs of Sale (% of value)

Building Costs
Office Building Costs (Gross, £ psm)
Office Building Professional Fees (% of building costs)
Building Contingencies (% of building costs)

Phasing	Date	Month
Start of Building Period		0
End of Building Period		0
Timing of Letting / Sale		0

Letting, Advertising & Sale fees
Letting fees (% of annual income)
Advertising fees (% of annual income)
Sale fees (% of sale price)

Developer's Return for risk / profit (% of value)

Figure 11.15 HCA DAT Non-residential property data entry screen

'GLA toolkit' style Scheme Results

Site Reference Details

Site Reference	0
Local Planning Authority	0
HCA Investment Partner	0

Site Details

Site Address	Dummy data
Scheme Description	Mixed

TOTAL NUMBER OF UNITS

Dwellings	40

DENSITY (per hectare)

Dwellings	49.4

AFFORDABLE UNITS

	Quantity	% all units
Total	0	0%
Social Rented	0	0%
Affordable Rent	0	0%
Shared Ownership	0	0%

REVENUES AND COSTS

Total Scheme Revenues		4,600,000
Total Scheme Costs	4,237,825	

Surplus/(Deficit) Present Value

Whole Scheme	330,933
Per net hectare	408,559
Per dwelling	8,273
Per market dwelling	8,273

Contribution to Revenue from

Market Housing		4,600,000
Affordable Housing		0
Social Rent	-	
Shared Ownership	-	
Affordable Rent	-	
Other Contributions		0
Non Residential Values		0

Alternative Site Value

Existing Use	1,500,000

Contribution to Costs from

Market Housing		1,062,353
Affordable Housing		0
Social Rent	-	
Shared Ownership	-	
Affordable Rent	-	
Other Construction costs		386,665
Planning Obligations		-
Fees		112,040
Non Residential Costs		-
Finance and Acquisition Costs		1,756,768
Developer's return for risk and profit		920,000

Figure 11.16 HCA DAT summary results

HCA DEVELOPMENT APPRAISAL TOOL

All values to nearest £

Dummy data
QUARTERLY CASH FLOW

	End of Quarter	Total	27-Jan-13	27-Apr-13	27-Jul-13	27-Oct-13	27-Jan-14	27-Apr-14	27-Ju
			Q0	Q1	Q2	Q3	Q4	Q5	
Month									
Site Acquisition Costs	Land Value	-1,500,000	-1,500,000						
	Agents Fees	-17,000	-17,000						
	Legal Fees	-17,000	-17,000						
	Stamp Duty	-85,000	-85,000						
	Other Acquisition Costs (£)								
Building Costs	Building Cost (Total)	-1,115,471		-134,524	-334,641	-334,641	-244,959	-66,706	
	Building Cost Fee (50% lump sum)	-56,774		-56,774					
	Building Cost (regular payments)	-56,774		-13,102	-13,102	-13,102	-13,102	-4,367	
	Car Parking Spaces	-20,000					-20,000		
Section 106 costs	Education								
	Sport & Recreation								
	Social Infrastructure								
	Public Realm								
	Affordable Housing								
	Transport								
	Highway								
	Health								
	Public Art								
	Flood work								
	Community Infrastructure Levy								
	Other Tariff								
	Other 1								
	Other 2								
External Works &	Site Preparation/Demolition	-50,000	-16,667	-33,333					
Infrastructure Costs (£)	Roads and Sewers	-75,000		-16,071	-16,071	-16,071	-16,071	-10,714	
	Services (Power, Water, Gas, Telco and	-25,000			-18,750	-6,250			
	Strategic Landscaping								
	Off Site Works								
	Public Open Space								
	Site Specific Sustainability Initiatives								
	Other 1								
	Other 2								
	Other 3								
Site Abnormals (£)	De-canting tenants								
	Decontamination	-50,000	-16,667	-33,333					
	Other								
	Other 2								
OMH Sales Fees	Sales Fees (OMH only):	-92,000					-13,333	-42,286	-28,952	-7
	Legal Fees (OMH only):	-40					-7	-19	-12	
	Initial Letting Fee ph1								

H ◄ ► H │ Input 5 -Res Costs │ Input 6 - Non Res │ GLAStyle Output │ Warnings │ **Output Qtrly CF** │ Output – Fill │ Notes & Memos │ C1 Mix │ C2 Value

Figure 11.17 HCA DAT cash flow output extract

The model then produces a summary output (Figure 11.16) and a series of cash flow outputs (Figure 11.17) that are not editable – i.e. they are outputs only.

Although this Excel spreadsheet model has a specific purpose, it has characteristics that are typical of many such 'professional' type Excel models. It does give the HCA the consistency it desires and, presuming that it has been constructed correctly and fully audited, will be reliable. That however, is one of its drawbacks; the model is opaque, the user cannot easily interrogate the model to see how the calculations are actually being done. The whole structure of the programme with its multiple screens for input and outputs is ungainly and relatively complex, although the model does prevent serious errors being saved. The model has attempted to be comprehensive but it is still inflexible' it does not allow the user to model specific features or complexities of their projects, perhaps forcing the user to make unrealistic assumptions that are not applicable to their circumstances.

Vulnerability of spreadsheet types to typical sources of error

Although the previous section was intended as a review of the broad categories of spreadsheets used in development feasibility studies, I have already started to identify areas where these models are sub-optimal in terms of vulnerability to mechanical or construction errors (as opposed to forecasting errors which is inherent in all appraisal models). A starting point is to identify the most likely types and sources of mechanical or construction errors in spreadsheets. It is difficult to be definitive about this and cite real world examples as few developers are willing to advertise or admit to their mistakes; however these sources of errors have been observed in practice and have been experienced personally by the authors:

(i)   *Errors due to time pressure* – many workplaces are high-pressure environments with appraisers having to do often complex work within a short timescale.

(ii) *Failure to properly audit the spreadsheet* – auditing can eliminate errors from spreadsheet but every creation of a spreadsheet item or change to a spreadsheet model requires an audit trail to be followed which costs the developer time – standardised models such as Estate Master DF do not need the same audit and therefore save the developer considerable time in checking the mechanics of the calculations.

(iii) *Incorrect modification of an existing spreadsheet model (and a presumed failure to audit)* – this is a common set of circumstances. Development projects are not static; there are always many changes from the initial appraisal, where many assumptions have to be made, to the final appraisal immediately prior to work on site.

(iv) *Application of an existing model to new development projects* – it is a natural thing when considerable time effort has been invested in the creation of a spreadsheet model to spread the cost (and save time) by applying and adapting the model for different projects. This opens up the possibility of modification errors as in (iii) above but also in the perpetuation of errors from earlier projects because the assumption will have been made that the applied model will have been audited and is error free on the earlier projects.

The *type one*, the simple residual models, and the *type two*, project specific, manually linked models, are particularly vulnerable to all of these sources of errors. This is particularly serious as these are the models probably the most commonly used in development appraisal practice. The models will probably have been audited for errors in their initial use but, as noted, modification over the project life is inevitable. Also, almost inevitably, once a model has been created, used and become familiar to a project team, it will almost certainly be applied to other projects. These spreadsheets are often used by smaller developers who will not have the luxury of extensive staffing, so it is likely that it is these organisations that will be under the greatest time pressure and will have little available to properly audit and re-audit for errors when changes are made.

They are particularly vulnerable to errors.

The *type three* models are more sophisticated and are usually designed to be applied across a range of projects rather than be constructed for one in particular. They should, therefore be more reliable. They are not immune from error however and their relative sophistication can lull the user into a false sense of security into believing that they are foolproof. In fact, their relative complexity is also an area of vulnerability; changes made are more difficult to do properly and the downstream effects of the alterations are often more difficult to appreciate. Development projects are so varied – different phasing, special cost distributions etc. – that it is very difficult, if not impossible, to design a spreadsheet template that will meet all variations. Ideally any changes should be made by the original model creator and properly audited; however the practicalities of working on development projects means that this cannot be assured.

The *type four* models should be more reliable because they have been professionally produced and tested and designed for a range of projects; however this does not make them issue free. For one thing is they are complex and opaque. A user may make a data entry error but not realise that they have erred. Another issue arises out of the fact that these models have a special purpose in their design. This tends to mean that they are limited in modelling other factors, elements that are probably very important to the developer. The HCA model is a case in point; although it does impose consistency across appraisals on developers there are elements, for example, the modelling of the non-residential elements in particular is a gross oversimplification and inadequate, yet the developer is forced to use it (and this itself is a potential source of error). Fundamentally too, for all the protection placed on the input screens, the model is still based on Excel. People with sufficient knowledge of Excel can turn this protection off and modify the underlying formulas – and this applies to most if not all of these sophisticated Excel models. Although this can produce the required tailoring to suit the requirements of a specific project, the complex opacity of these models makes them equally vulnerable to errors.

A final point in this argument is to consider the increasing investment in time and money in these models. The advantages of the simple, project-specific models is that they are relatively cheap and flexible, if very vulnerable to errors. This situation can be improved by investing in a professionally produced system but this still produces a sub-optimal result yet the cost of producing them almost certainly outstrips the cost of purchasing licences of tried and tested, consistent and reliable, yet flexible proprietary systems.

Conclusions on the use of Excel in development appraisal

Development appraisal is a very different animal from the more predictable world of standing investments, and, indeed, different from all of the environments mentioned in the correspondence above. Different from designing aircraft, modelling sophisticated financial instruments and multi-million dollar/pound/yen/euro standing investments? Absolutely so.

The key differences are volatility, risk, time pressure and heterogeneity. Although the mathematics and calculations in a development appraisal ARE relatively simple, because of the geared nature of the development calculation (i.e. you are calculating the margin of return) the results are extremely volatile, as we have seen from the section on sensitivity. Only minor changes in assumptions make huge changes in the output of the appraisal. A slight mistake can have a major impact.

However, what about the transparency of Excel, surely you can just audit your models to find the errors? Well, in theory yes but in practice two factors conspire to prevent that. The first is that most initial appraisals are done under a degree of time pressure; there often is simply not the time to run a comprehensive audit. The second is down to a characteristic of humans: the inability to see our own mistakes. It is a fact that most Excel models are self-created by the people carrying out the appraisal. Anyone who writes anything knows that it is essential to get a second pair of eyes to look at anything important; this second viewer will instantly spot glaring mistakes the creator has missed. Most appraisers do not have this luxury.

Surely, it might be argued, as models are developed over time, these mistakes will be picked up and removed? This would certainly be the case with appraising standing investments where many components will be consistent and repeated. It is NOT true of developments. Each development tends to be very different, even if it is only in the duration of the project. Whatever, every Excel model will inevitably have to be changed from project to project. As soon as you have change you have the potential for error – and the effect of every error is magnified because of the volatility of appraisal yet you don't have time to track down that error . . . etc., etc.

This is why I am such a strong advocate of tried, tested and reliable proprietary packages. They are designed for the job, they have uniformity and consistency meaning that you are actually prompted to put in all of the components you need (another potential source of error). They even have the ability to import and export to Excel if you need additional flexibility or special calculations.

Excel is a fabulous tool, it has immense power; however the special nature of development means that in using self-created models you are playing with a ticking time bomb that will one day blow up in the user's face with potentially devastating financial consequences.

12 Modelling development financial feasibility in Argus Developer – software introduction and case studies

An overview of Argus Developer

History

Argus Developer started its life as Circle Developer.

Circle was the company founded by Adrian Katz in 1990 initially with Developer as its sole product, although this was soon followed by Circle Investor valuation and investment appraisal software. The original DOS version of Developer sold well into what was then a fairly competitive and crowded market place. In the mid-late 1990s it was the launch of the heavily revised windows version of Developer, initially called Visual Developer, that established the company on a path that rapidly saw them dominate the market. From this initial success the program has continued to be developed though the principle structure and layout remains largely unchanged.

In 2006 there were a number of key events. Firstly Circle was bought by its larger American rival, the Houston-based Realm Business Solutions. Subsequently the whole group was rebranded under the Argus Software banner and Circle Developer became Argus Developer. In the same year a major upgrade of the program was launched, Developer 3, which saw improved functionality of the main program and the addition of a new and very powerful structured finance module. There have subsequently been two more version changes, where additional functionality in regard to residential development appraisal and operated assets such as golf courses, hotels and marinas were added, though the core way the program works and calculations run have remained constant. Someone familiar with Circle Visual Developer would quickly feel comfortable with Argus Developer version 5 or 6.

An introduction to the program

Note this book uses Developer version 5 to illustrate the models. Argus have recently launched Developer version 6. There are only minor differences between versions 4, 5 and 6 and the bulk of the material illustrated in this book would be equally familiar to users of versions 2 and 3.

Argus Developer is primarily a cash flow calculation tool. Data entered into the program is placed within either a pre-set or user defined timeframe and calculation is made using discounting based on the interest assumptions entered. One of the principal outputs is a summary laid out as a traditional residual appraisal but this is simply to present the data in a familiar layout to the user and any third parties. No calculation is undertaken using the residual sheet.

Argus Developer has four main screens or tabs in which data is entered and where the output/results of the calculation are displayed. These tabs are project, definition, cash flow and summary. There is also a toolbar and drop down menu allowing various functions of the program to be accessed.

The project tab

Project is where the address details of the project are entered and also where the main calculation assumptions are made. As we shall see these are broad, global assumptions that can be changed or overridden at any point during the calculation. They can be used to make broad assumptions where a quick calculation is required.

The project tab includes two shortcut buttons that take the user to key assumption elements. There is a third button, for structured finance. This is an add on module enabling the modelling of more complex financing such as partnerships and different sources and costs of finance. Unless this module option is taken this option is greyed out.

The first is assumptions for calculations (Figure 12.2).

Developer is preloaded with a template – essentially a default set of assumptions – that allows rapid appraisals to be carried out. These assumptions can

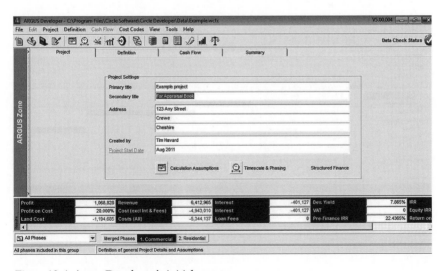

Figure 12.1 Argus Developer's initial screen

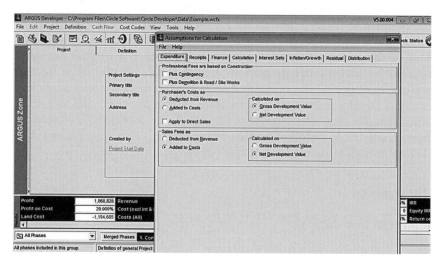

Figure 12.2 The assumptions for calculations

be altered or just left at the default settings (and there is the facility for user created templates to be held on the system). For all appraisals, however, two areas must be visited by the appraiser; the interest sets sub-tab and the residual sub-tab which sets the mode of calculation. These will be covered in detail in the worked example section below. You will observe that there are eight individual tabs in the assumption for calculation section.

The second shortcut button takes the user to an important set of assumptions, the timescale and phasing section (Figure 12.3).

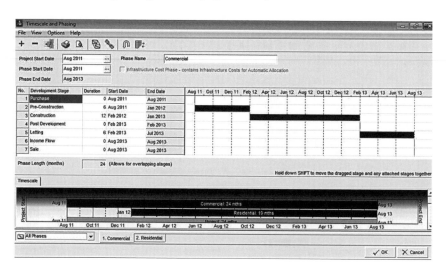

Figure 12.3 The timescale and phasing screen

The timescale assumptions are an essential part of the calculation and assumptions must be made here. The development is broken down into seven stages. The user does not have to define timescales for each of these stages, only those appropriate for the scheme.

Two things are important to know; firstly, the template of the program comes preloaded which assigns activities and events to each of these stages, for example any new construction work will be assigned to the construction phase with an automatic s-curve distribution. Demolition work will be assumed to be a one-off event at the beginning of the construction phase and so on. The second thing to note, however, is that each individual event can be individually timed and distributed to suit the particular project. These pre-set assumptions can all be overridden.

The definition tab

The definition tab is the primary (but not the only) point for data entry. It is divided into groups of associated cost and income fields (Figure 12.4).

Most of the data entry can be carried out by typing values and assumptions straight into the relevant box but there are some areas where it is necessary to drill down behind the box to make detailed entries. The principal boxes where this occurs are the four grouped boxes in the upper left part of the definitions tab (Figure 12.5).

This is where the main value and construction cost elements are calculated. The capitalised rent box is designed mainly for income producing commercial property whilst the three other boxes are for different types of residential development calculation.

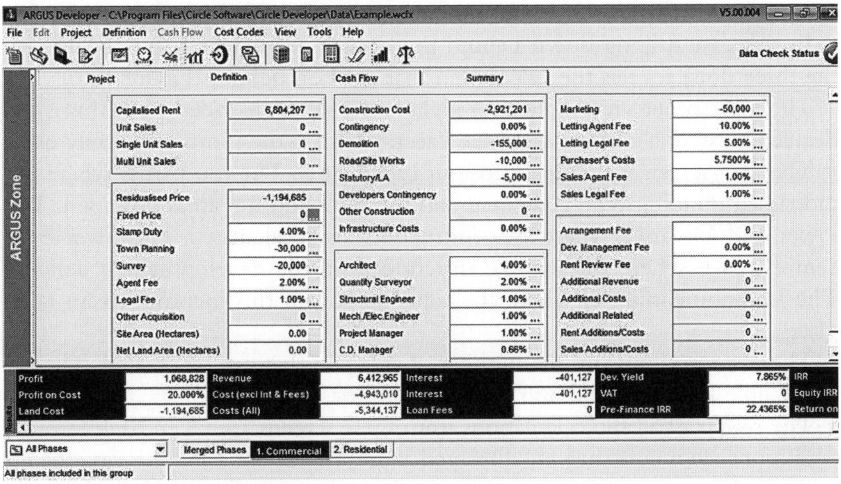

Figure 12.4 The definition tab

Capitalised Rent	6,804,207 ...
Unit Sales	0 ...
Single Unit Sales	0 ...
Multi Unit Sales	0 ...

Figure 12.5 Detail of definition tab – the grouped project type shortcuts

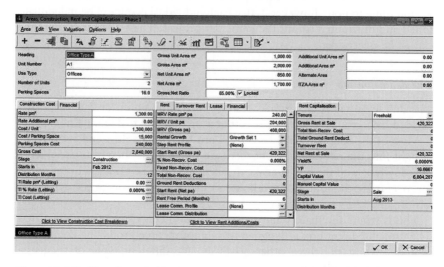

Figure 12.6 Drilling down behind to reveal the capitalisation of income screens

In this case drilling down behind the capitalised rent box (by clicking on the three dots) reveals the calculation that goes on behind (Figure 12.6).

There are other view options – including a schedule view – but this is the detailed view of a commercial area screen which is the most commonly used.

The area screen is divided into four broad parts. The top part is where the individual building type is defined and the built and net areas provided. The lower-mid-left area is where the construction cost is calculated. The lower-mid-centre area is where the income structure from the development is defined. The lower-mid-right is where the capitalisation of this income stream takes place.

There are no limits on the number of area tabs that can be created. The tab can handle both freehold and leasehold interests.

The results of all the calculations from these screens are taken back into the definitions tab where they are displayed as greyed out boxes, in this case against capitalised rent and the construction cost boxes (refer to Figure 12.4). An entry here is also reflected in the cash flow and the summary outcome sheet.

The remaining boxes in the definitions tab allow all elements of the development to be costed or valued and these elements are then placed in the relevant timeframe. This can be done using the preloaded template or else each individual cost and revenue element can be manually determined.

This can be demonstrated with the demolition element of this appraisal. This box is greyed out illustrating that there is more than one element to it (Figure 12.7). Drilling down into the box (using the button indicated by the three dots) reveals the detail of the calculation (Figure 12.8).

As can be observed there are two elements to the demolition. These extra elements have been created by drilling down into this section and adding new lines using the + symbol. As it happens these items are both stand alone cost items but the program allows a number of different calculation options such as relating the cost to another element or elements or relating the cost to an area (Figure 12.9).

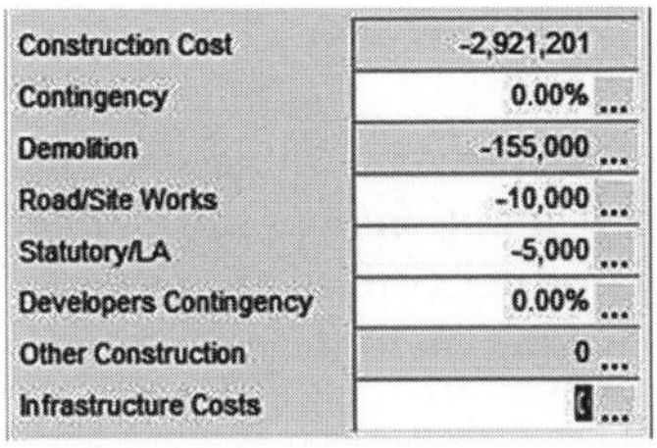

Construction Cost	-2,921,201
Contingency	0.00% ...
Demolition	-155,000 ...
Road/Site Works	-10,000 ...
Statutory/LA	-5,000 ...
Developers Contingency	0.00% ...
Other Construction	0 ...
Infrastructure Costs	...

Figure 12.7 Detail of definition tab – construction cost and related fields

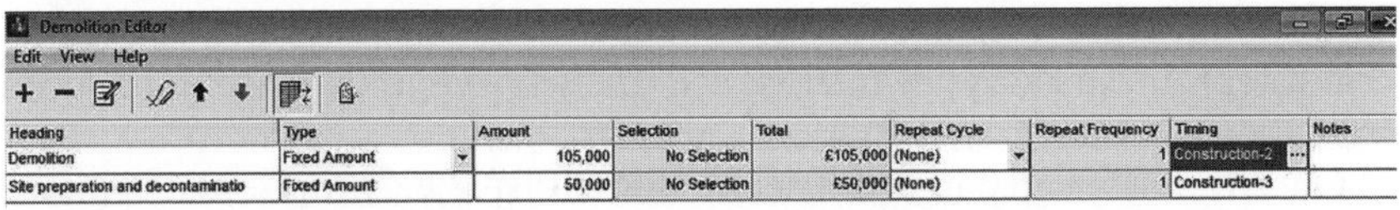

Figure 12.8 Drilling down behind the demolition section

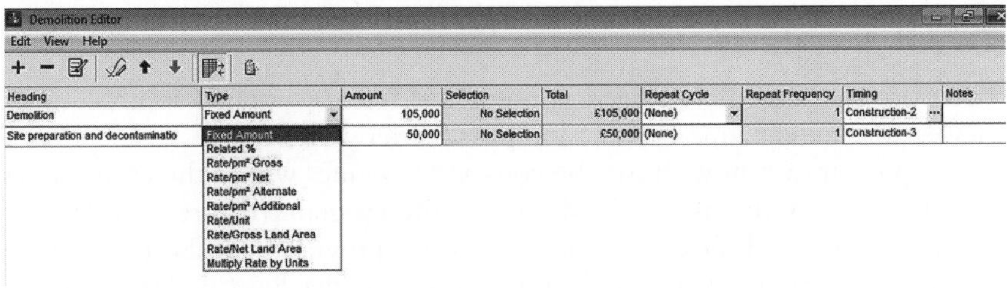

Figure 12.9 Detail of demolition screen showing degree of variability/control
possible with each element

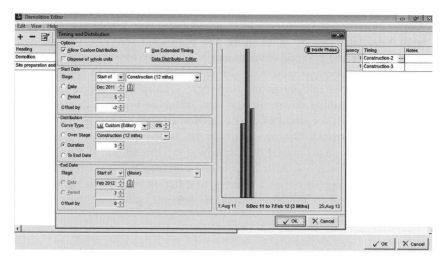

Figure 12.10 Adjusting the timing and distribution of an expenditure

As to timing, as noted, Developer assigns costs and revenues in accordance with the pre-existing template. In the case of the demolition item this is timed by default as a single cost item to the beginning of the construction stage of the project. This timing can be over-ridden by clicking on to the relevant row in the timing column (Figure 12.10).

Any distribution required can be modelled.

This, in essence, is the power of Developer as a program. It can be used quickly if a quick feasibility study is required, the template allowing a rapid calculation of viability using reasonable approximations. It can, however, be used as a finer tool with more in–depth assumptions. The program is also highly transparent. It can always be interrogated to discover exactly what assumptions have been made.

Once the data entry into the definitions tab have been made the appraisal is essentially complete. The results of the calculation can be seen in the results bar at the bottom of the screen which gives an instant summary of the calculation and also in the cash flow and summary tabs of the program.

The cash flow tab

There are two cash flow screens, project and finance, each of which can be viewed in different time cycles; monthly, quarterly etc.

The project cash flow shows the costs and revenues within the time–frames defined either by the user or by default by the program (Figure 12.11).

Although the cash flow is an outcome of the program it is also the primary calculation engine and the distribution of any of the lines within it can be reviewed and adjusted if required. This is illustrated with the office construction line in Figures 12.12 and 12.13.

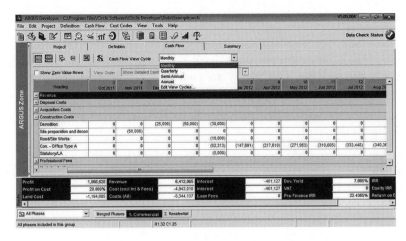

Figure 12.11 The cash flow tab – project view

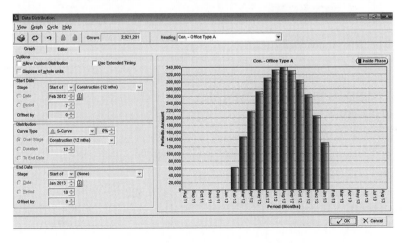

Figure 12.12 The cash flow tab – adjustments

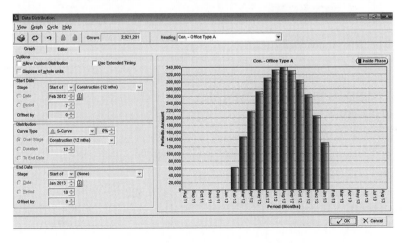

Figure 12.13 Accessing data distribution in the cash flow tab

New lines and new cost/revenue elements can be added directly into the cash flow. These automatically write back to the definitions tab keeping the records up to date. This ability to control and fine tune underlines Developer's power.

The summary tab

The final tab is the summary screen. As noted above this is laid out as a traditional residual appraisal and gives a quick, concise and clear overview of the development (Figure 12.14).

The summary can be printed directly from this tab or converted to PDF format for easy transmission. It can also be printed via the reporting module and exported into Word (see example below).

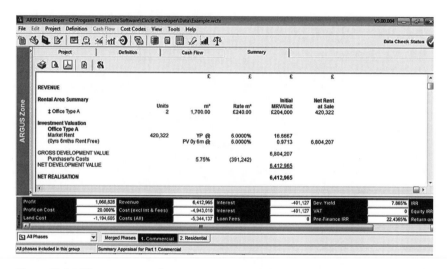

Figure 12.14 The summary tab

CONDENSED APPRAISAL SUMMARY
   Example project
   For Appraisal Book

Condensed Summary Appraisal for Part 1 Commercial

INCOME

Annual Rental Income	420,322	
Net Capital Value		6,804,207
Less Purchaser's Costs		(391,242)
Net Realisation		6,412,965

OUTLAY
    Acquisition
        Site Purchase Cost   1,194,685
        Site Purchase Fees   133,628
        Total Purchase Cost   1,328,313

Construction
        Construction Costs   3,091,201
        Professional Fees   282,188
        Total Construction   3,373,389

Marketing/Letting
        Marketing   50,000
        Letting   63,048

Disposal
        Sales Costs   128,259

Finance
        Project Length   25 months
        Multiple Finance Rates Used (See Assumptions)
        Site Finance   132,101
        Construction Finance   100,945
        Void Finance   168,081
        Total Finance   401,127

Total Expenditure   5,344,137

Profit   1,068,828

Performance Measures
        Profit on Cost%   20.00%
        Profit on GDV%   15.71%
        Profit on NDV%   16.67%
        Development Yield%   7.87%
          (on Rent)
        Equivalent Yield%   6.00%
          (Nominal)
        Equivalent Yield% (True)   6.23%

        IRR   22.44%

        Rent Cover   2 yrs 7 mths
        Profit Erosion   2 yrs 8 mths
          (finance rate 7.000%)

Other features

The program has a number of other features including a full reporting module
(Figure 12.15).

It also has graphical elements that can be exported into the reports (Figure 12.16).

Very importantly, given the nature of development, the program also has a sensitivity analysis module (Figure 12.17).

This is a powerful tool, although it is perhaps the least well developed and, from my observation, the least well used in practice of the elements of the program. We will examine the best use of this module later in the book.

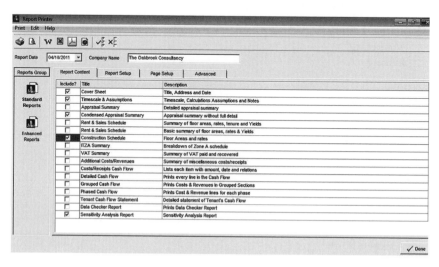

Figure 12.15 The reporting module – note export options

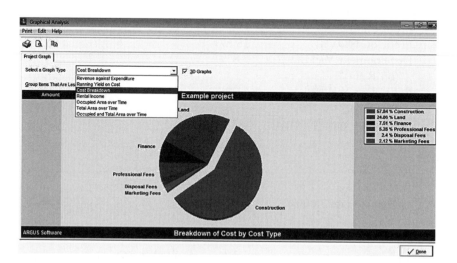

Figure 12.16 Graphical analysis

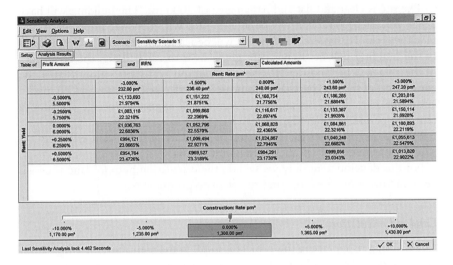

Figure 12.17 Sensitivity analysis module

This section is intended to give a broad overview of the program. A more in-depth appreciation of its use and power will be given in the worked sections below. I think it can be seen already, however, that Argus Developer is a very powerful, very comprehensive and complete appraisal tool. It has very distinct advantages over self-created Excel models – it is reliable, flexible and consistent.

The best way of appreciating the qualities of the program is with case studies, and these begin in the next section.

Case studies – commercial development appraisal

The appraisal of commercial development projects used to be the primary activity for the surveyor involved in valuation and development consultancies. However, as noted in the first section, this activity has been affected by the changes in the development market and practitioners will probably appraise far more residential development schemes and mixed-use projects than pure commercial schemes. Notwithstanding this, this sector of the market is still very important and this type of purely commercial appraisal is still very common.

We will look at pure commercial schemes in this section, starting with simple projects and then adding more complexity.

Argus Developer case study one – simple one building/one type/one phase commercial project

The initial project we will be appraising here is the development of a speculative single, three-storey office building on a greenfield site.

The site is clear and flat and has an area of 3000 m². The building will have a plan area of 1000 m² and will be steel-framed with concrete floors, brick and block walls and casement windows. It will be naturally ventilated. The grounds of the building will be 80 per cent given over to car parking, the remainder of the area will be landscaped. The construction will start 3 months after the completion of the land purchase. Construction will take 6 months and a 6 month letting up period will be allowed. Tenants will expect to receive a 3 month rent free period and sign 5-year leases.

Current interest rates for development loans are at 6 per cent effective.

The developer will sell the building on letting.

For this development only we will do the calculation both for land residual, i.e. Land price or valuation, and to calculate profitability.

Land bid calculation

The first step in appraising this project is to fill in the identification information on the project screen (Figure 12.18) and to create a save file for the project. Developer does not automatically save the data, although like Excel it has got a recovery save function and you are prompted to save on closing the program.

Once the address data has been entered we can then save the project (Figure 12.19). It is recommended that the project file is saved regularly as data is entered.

The next step is to review the assumptions for calculations and to make the necessary entries for interest costs and mode of calculation.

The assumption options are accessed either from the drop down menu, the button on the icon ribbon or else the button on the project tab itself. All take the user to the same screen.

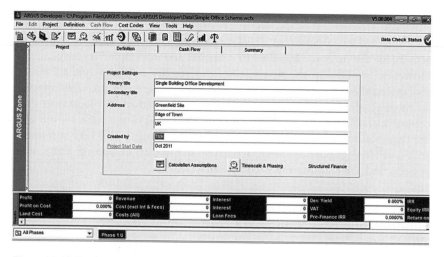

Figure 12.18 Project screen

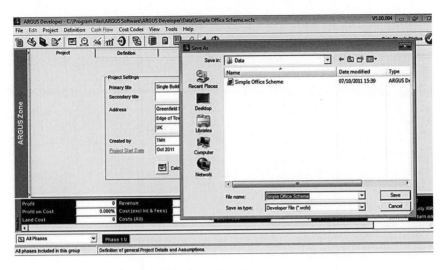

Figure 12.19 Saving the project in Argus Developer

As I have already noted in the introduction, Developer has a pre-existing template preloaded. The template uses reasonable assumptions derived from market experience and are applicable to many projects. All of the pre-set assumptions in the template can be adjusted by the user.

Over the next few pages I show the assumptions for each individual sub-tab and will outline what each of the options are. This will not be done in this detail in the later appraisals in this book.

The expenditure assumptions (Figure 12.20) contained within the template are, as with other aspects, those which are most commonly used in the UK development market, with the most popular alternative options available by checking the appropriate button. This section deals with the calculation of professional fees, the treatment of purchaser's costs and sale fees.

The default settings for professional fees is for them to be calculated against the build cost, which is the more common option in projects, but this definition can be widened if the professional team has wider responsibilities. Again it will be stressed that this is only the global set up; any individual item of expenditure or receipts in Developer can be individually defined later in the program. Here the default is to set professional fees as percentage of the costs expended but the user can define the professional fees as, say, a lump sum if required.

Similar switching options are available for the treatment of (incoming) purchaser's costs and sale fees (i.e. the developer's cost of disposal). With the former the default is to deduct the costs from the sale receipts; however, for lending and performance reasons some developers prefer to add the purchasers costs to the development costs.

The next of the set up assumptions deal with receipts.

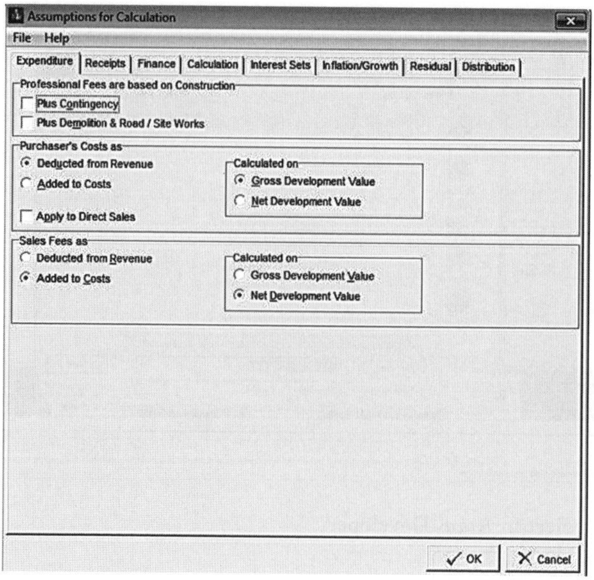

Figure 12.20 Expenditure assumptions

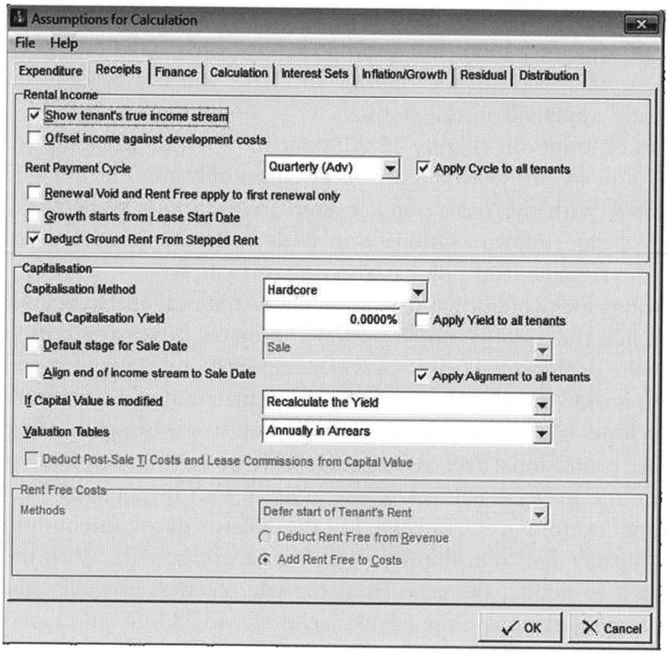

Figure 12.21 Receipts standard assumptions

The receipts section (Figure 12.21) is similarly set up with the industry norms (for the UK market at least) in mind. It is divided into two sections, the first dealing with the rental income stream, the second with the capitalisation assumptions.

There is a third, greyed out section that allows an alternative treatment of any rent free period in the development calculation. This section is made live by unchecking 'Show tenants true income stream' in the rental income section. This would normally be done to switch the rent free period to a development cost rather than as a deduction from capital value, a common choice for developers as it gives them financing advantages by maximising the projected end value of the scheme.

The top part of the screen deals with the income stream that the development produces. Options available include switching the income cycle from the default quarterly in advance setting, the norm for UK commercial leases. For residential investments, for example, it would be more appropriate to move to a monthly income pattern. Note, however, that these changes will only have a significant impact if there is a substantial 'hold' period envisaged.

The lower half of the tab deals with the capitalisation of this income stream. The capitalisation of income is normally calculated and defined within the area sheets accessed from the definitions tab (see below), however it is possible to set a global capitalisation rate. In practice, this is a rarely used option.

The third of the assumption screen tabs deals with finance.

Argus Developer's base version, which all the models presented in this book are constructed in, has a relatively simple finance module (Figure 12.22), although it can be used in quite a sophisticated way with the ability to set multiple rates and to tie any loan into any element of the project. To do proper debt/equity analysis however the structured finance module must be purchased. This is a very sophisticated tool allowing the modelling of complex financing and joint venture arrangements.

The standard module allows a global change to be made to the compounding and charging periods and also as to whether the interest is nominal or effective – i.e. taking account of the number of charging periods in the year.

There are four switching options at the bottom of this screen which allows fine tuning of the calculation. Here the normal set-up is being used; the only option I feel that should be made is to switch the quarterly compounding period to monthly as this is far more common in today's market (Figure 12.23).

Argus Developer allows the user to model changes in costs (inflation) and values (growth). A number of modelled sets can be created. The mere creation of these sets will not, in themselves, make any changes to the appraisal. The assumptions have to be manually ascribed to elements in the appraisal (using the drill down function as outlined in the software summary). This is sensible as these assumptions can make the appraisal extremely volatile.

The next section to consider is concerning the distribution of payments and receipts.

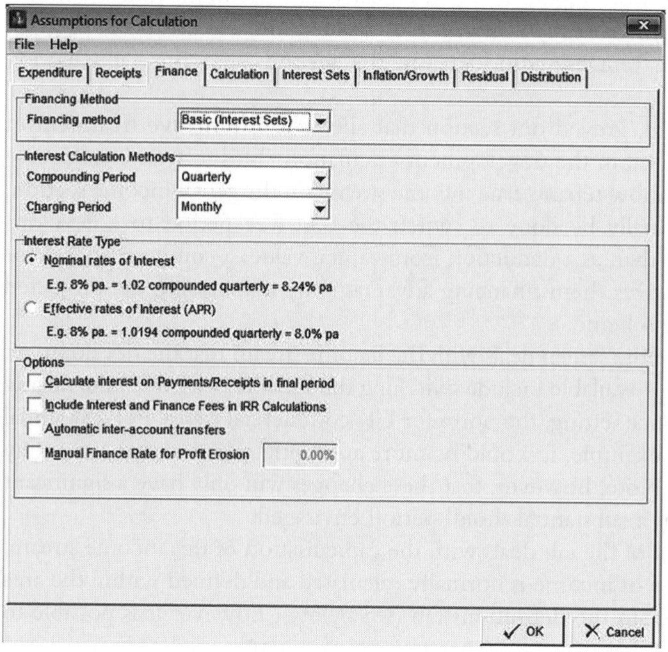

Figure 12.22 Finance standard assumptions

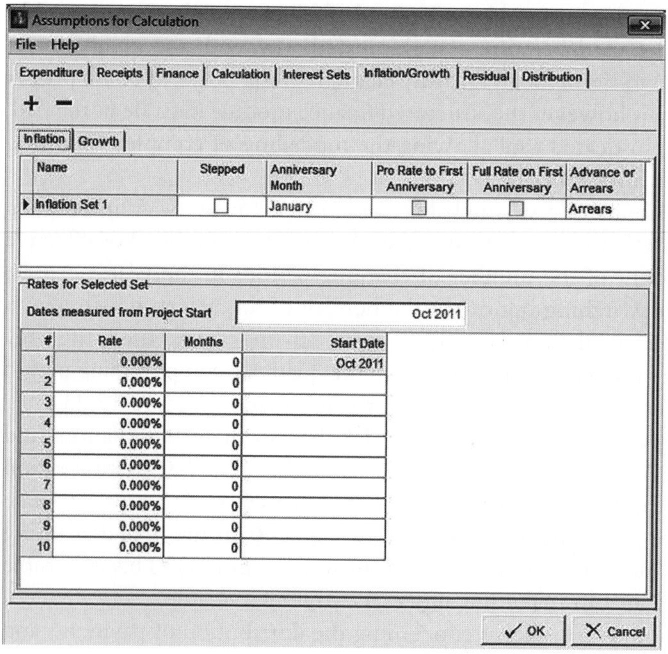

Figure 12.23 Inflation and growth standard assumptions

As with all aspects of inputs into Developer, all distributions can be modelled manually, however for convenience there are the most popular distribution options within the assumptions for calculation, allowing quick appraisals to be constructed (Figure 12.24). The template has a simple commercial project in mind; for a residential scheme, with multiple sales over a longer time period a more appropriate distribution can be selected, though this would only normally be done for a basic, initial feasibility study.

The final section to look at prior to dealing with the sections where changes must be made is the assumptions for calculations, i.e. the mechanics of how aspects of the timing of cash flows in and out are treated, and how IRR etc. is calculated (Figure 12.25). This tab contains a lot of information, reflecting the transparency of the program to the user, showing just how it is set up to calculate and allowing fine tuning to be made.

I am content to let the standard assumptions run for this appraisal, however there are two entries that must be made, and these are found on the tabs that I have not yet reviewed. The items we must address in this appraisal are the interest sets and the mode of calculation, found on the residual tab.

Developer needs at least one entry in the interest sets section (Figure 12.26). Multiple sets can be created but the program, by default, only reads the first one on the list, any subsequent ones have to be allocated to specific items in the definition tab. You will note that there are two columns in the basic interest

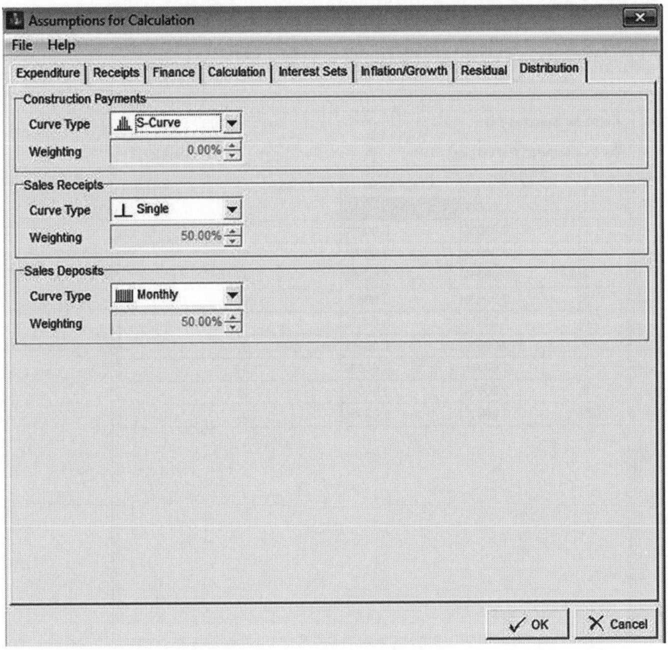

Figure 12.24 Distribution standard assumptions

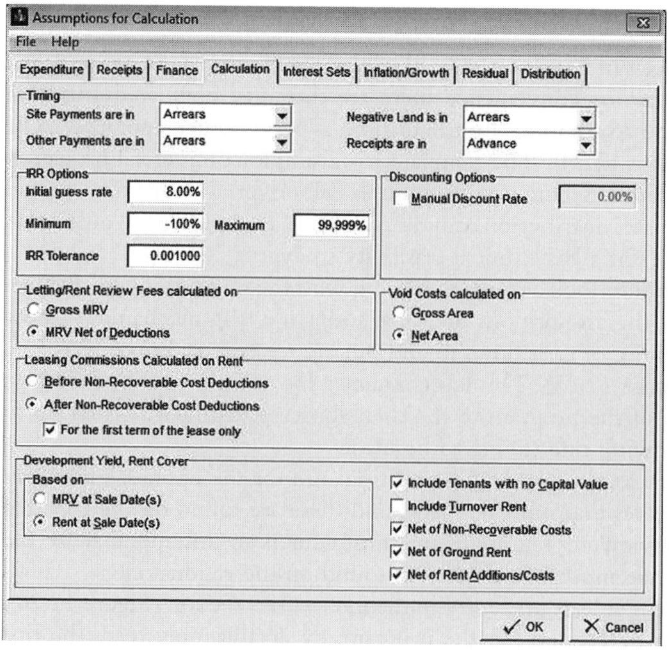

Figure 12.25 Calculation standard assumptions

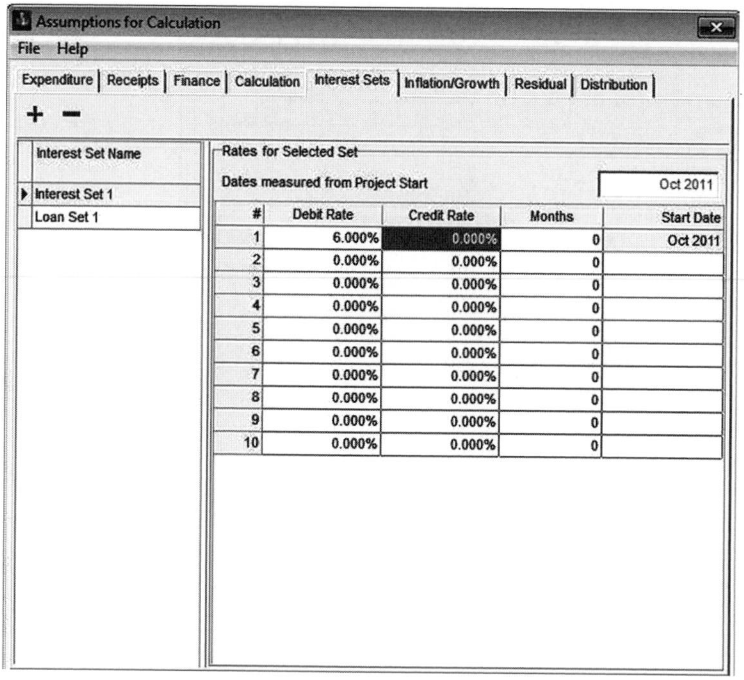

Figure 12.26 Setting up the interest sets for the development

sets module; the debit column is the rate of interest on borrowed money, the credit deals with interest earned on project surpluses. The only requirement is to make an entry into the debit column as here. Note also the column reading months. This column allows the modelling of variable interest rates, so it is possible to have, say 6 per cent applying for 12 months and then 5.75 per cent thereafter etc. It is advised that when these structured interest sets are created, the final row should have a rate and a '0' in the months column. This has the meaning of 'and thereafter' in Developer and avoids the interest rate charged on the development dropping to zero after the modelled structured element has expired.

The use of structured and multiple interest sets will be covered in later sections/examples below.

The final area which needs to be addressed at the assumptions stage is the mode of calculation.

As noted in the introduction, development appraisal/feasibility studies are undertaken to answer one of two alternative questions; either, what amount can be bid for the land whilst meeting the required target return or, where the cost of the land is known, what profit (or loss) will the project generate? Quite frequently a developer may well run the calculation in one mode in order to inform their bid to secure the site, and then later in the other mode as a development monitoring and project design tool.

Note that Developer also has a third option, a part residual land value/part profits calculation mode used only in special circumstances.

Because we are doing a land residual calculation we need to change the setup of the program as the default setting is to calculate development profitability against a known land value (Figure 12.27).

Clicking the residualised land value only option (Figure 12.28) allows assumptions to be made regarding the required developer's profit, the element which has to be fixed as a target so that the unknown element in the development appraisal, the land value, can be calculated.

Developer has six options for calculating (or perhaps, more accurately, definition) developer's profit (Figure 12.29). The most commonly used is return on cost, the total sums actually expended on the project. The alternatives are more rarely used. Profit on GDV and Profit on NDV are profits related to the end value of the scheme, either gross development, which excludes the costs involved in selling the building, or net development value which includes these costs. IRR is internal rate of return, the standard tool for measuring the returns from an investment. Development is a form of investment so sometimes using this measure is more appropriate. Development yield allows the developer to define a yield, or return, that the scheme must produce in terms of income on completion relative to the costs of completing it. Similarly, developers simply want to make a particular sum, and the program is flexible enough to allow this option to be calculated.

In this example we will stick to return on cost as this is the most common benchmark used. We will use a rate of 20 per cent which may appear high to

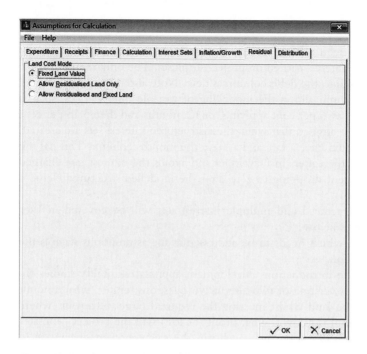

Figure 12.27 The standard residual tab setup

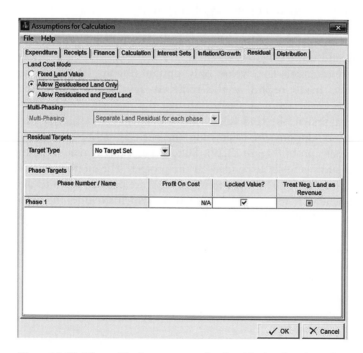

Figure 12.28 The residual screen once land residual only selected

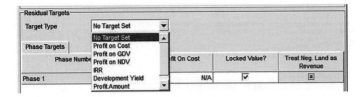

Figure 12.29 The alternative targets for profit

the layman but is a normal benchmark for a commercial scheme. We shall see later that the geared nature of appraisal (the relationship between end value and input costs) means that development is a high-risk endeavour and such wide margins are necessary to cover the risk elements.

Note that the locked value and treat negative land value as revenue should remain unchecked at this stage (Figure 12.30).

Once this screen is closed a warning is flashed up that the mode of calculation has been changed. Often too the data checker reports that no land value can be calculated (Figure 12.31). This is perfectly normal at this stage and can be ignored.

The next element we must deal with is timescale. The timescale and phasing section of the program is intended to set up the broad timings only. The cash

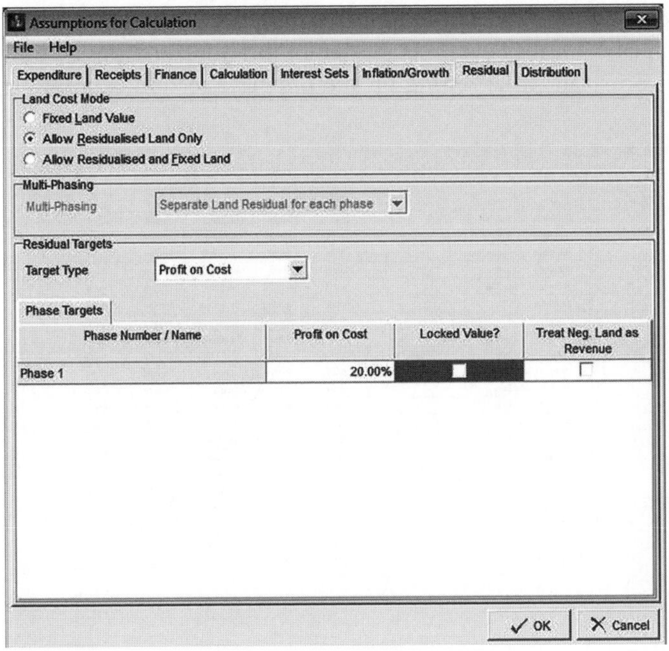

Figure 12.30 The residual screen final setup

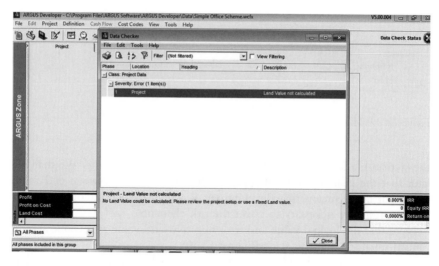

Figure 12.31 Data check warning screen which can be ignored at this stage

flow template preloaded into development has the elements of the project associated with certain time elements and setting up the timescales allows these to work. These assumptions can be viewed by selecting File/Administration/View Cash Flow Templates from the drop down menus (Figure 12.32).

Developer allows separate assumptions for different user defined projects to be saved as a template. This is beyond the scope of this book and this section is only included to explain what setting up the timescale section is doing. It

	Description	Brief Descr.	Order	Category	Int. Grp	Sign	Development Stage	Related Groups	Distribution	Distribution Curve	Relation Type	Relation Basis	Component Type	Dec.	k In
1	Site Area	Undef	0	0	2	Cost	(None)		Start	Single	Unrelated	Straight	Fixed	0	
2	Realisation	Realis	1	1	2	Revenue	Sale		Start	Single	Unrelated	Straight	Area	0	
3	Equity Profit	EquityProf	1	82	2	Revenue	(None)		Start	Single	Unrelated	Straight	Fixed	0	
4	Additional Revenue	OtherRev	1	5	2	Revenue	Phase Start		Start	Single	Unrelated	Straight	Fixed	0	
5	Existing Income pa	Exist Inc	1	6	2	Revenue	Sale		Start	Single	Unrelated	Straight	Fixed	0	
6	Rent History	MRV History	1	78	2	Revenue	Letting		Start	Related	Unrelated	Straight	Area	0	
7	Annual Rent	MRV	1	3	2	Revenue	Sale		Start	Related	Unrelated	Straight	Area	0	
8	Tenant Rent Flow	RentFlow	1	7	2	Revenue	Income Flow		Start	Related	Unrelated	Straight	Area	0	
9	Turnover Rent	TOverRent	1	83	2	Revenue	Letting		Start	Related	Unrelated	Straight	Area	0	
10	Rent Free Costs	RentFree	1	61	2	Cost	Sale	003	Start	Related	Line 2 Line	Straight	Related	0	
11	Geared Ground Rent	Gr Rent %	1	15	2	Cost red. Rev	Sale		Start	Related	Unrelated	Straight	Fixed	0	
12	Fixed Ground Rent	Gr Rent F	1	23	2	Cost red. Rev	Sale		Start	Single	Unrelated	Straight	Fixed	0	
13	Capitalised Rent	Cap Rent	1	4	2	Revenue	Sale		Start	Single	Unrelated	Straight	Area	0	
14	Unit Sales	Sales	1	2	2	Revenue	Sale		Start	Single	Unrelated	Straight	Area	0	
15	Sales Deposits	Sales Depos	1	88	2	Revenue	Sale		Start	Related	Unrelated	Straight	Area	0	
16	Rent Additions/Costs	AdtRntRevC	1	89	2	Revenue	Letting		Start	Single	Unrelated	Straight	Related	2	
17	Additional Rent Rev/Co	AdRRevCos	1	115	2	Revenue	Sale		Start	Single	Unrelated	Straight	Related	2	
18	Sales Additions/Costs	AdtSlsRevC	1	90	2	Revenue	Sale		Start	Single	Unrelated	Straight	Related	2	
19	Op. Asset Rev/Expense	OpAssetRev	1	97	2	Revenue	Income Flow		Start	Related	Unrelated	Straight	Area	0	
20	Op. Asset Rev./Expens	OpAssetRev	1	98	2	Revenue	Sale		Start	Related	Unrelated	Straight	Related	2	
21	Disposal Proceeds	Disp. Proc	1	72	2	Revenue	Phase Start		Start	Related	Unrelated	Straight	Fixed	0	
22	Purchaser's Costs	Purch Cost	1	8	2	Cost red. Rev	Sale	004	Start	Related	Line 2 Line	Straight	Related	4	
23	Sales Agent Fee	Sale Agent	1	9	2	Cost	Sale	004/003/008/115	Start	Related	Line 2 Line	Straight	Related	2	

☑ Fix Heading Column when scrolling   ✓ Done   ✗ Cancel   Print   Preview

Figure 12.32 Extract from the default cash flow template showing associations and timings

has been stated before but is worth reiterating that the assumptions made for any individual cost or value element can be overridden by the user.

Pressing the shortcut button (Figure 12.33) brings up the setup screen for timescale and phasing (Figure 12.34).

Developer uses seven broad development stages. These cannot be increased, although the names of each stage can be changed. Not all the stages need to have a time set against them, only those which are appropriate.

Some users on first introduction to this section of the program are surprised that only seven stages are defined and that this seems to be oversimplifying and restrictive. In practice, however, it is not. As noted, these are just broad-brush timings, any individual item can be separately timed, the stages can overlap, and the ability to add phases which can be consecutive or concurrent (or, indeed, both) means that the program can deal with the vast majority (if not all) of specific timing requirements. The Timescale and Phasing tab is essentially a visualisation tool albeit with key expenditure and receipts tied to it via the default template.

Here we will be using only three of the sections: the 3 month planning period, the 6 month construction phase and the 6 month letting up phase. Note the

Figure 12.33 The timescale and phasing shortcut

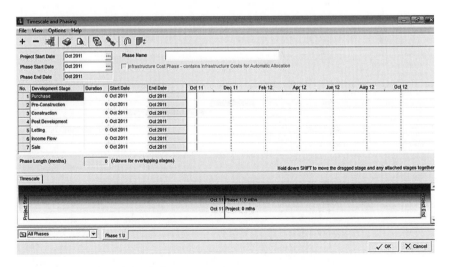

Figure 12.34 Timescale and phasing screen

rent free period will be dealt with later. Data on timescales for projects can be obtained from a number of sources including the client, architect or quantity surveyor, or from experience of other, similar projects. Sometime, however, particularly at the feasibility stage of a project, a surveyor may have to simply form a reasoned judgement on timings (Figure 12.35).

For completeness, the purchase stage allows for a delayed, or staged land purchase, post development is designed for a snagging and fitting out period best suited to the development of a building for a known end user, income flow is for retained investments and sale allows the modelling of a longer sale period (usually) for multiple units as is found with residential schemes. As can be seen, none are really appropriate for our simple speculative scheme produced by a developer-trader.

Developer is now set up to receive data. We can therefore move onto the definitions tab, remembering to save our project (Figure 12.36).

The starting point for our project on this screen is to drill behind the capitalised rent box, double clicking the three dots to open up the area screen (Figure 12.37). This is where the majority of the cost and value assumptions will be made.

The commercial area screen can be viewed in different modes in the program but the most common view used is the detailed screen shown in Figure 12.38. The views can be toggled using the drill (detailed view) and spreadsheet (schedule view) icons.

The area screen is divided into four functional areas. The top part of the screen defines the use type (used for the reporting and sensitivity elements in the main) and allows the setting of areas, both the gross area (i.e. the built area) and the net area (the net lettable area). The latter can be entered directly or

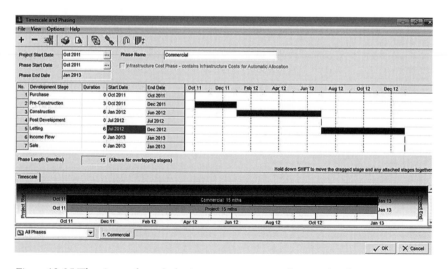

Figure 12.35 The timescale and phasing screen as set up for our development

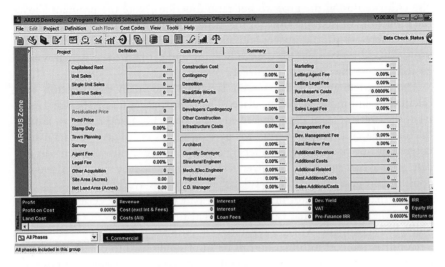

Figure 12.36 The definitions tab for our project

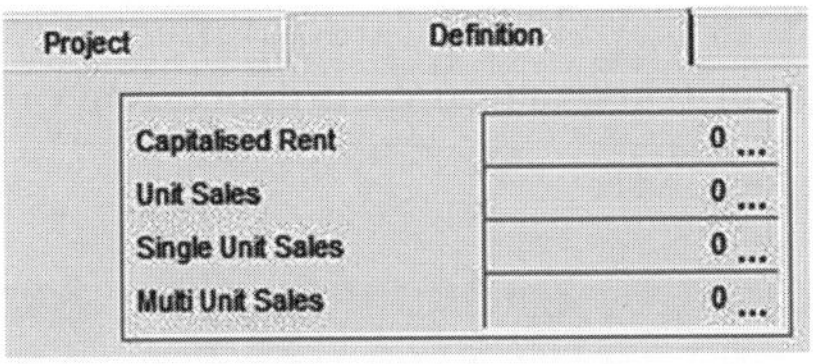

Figure 12.37 The shortcuts to the area screens

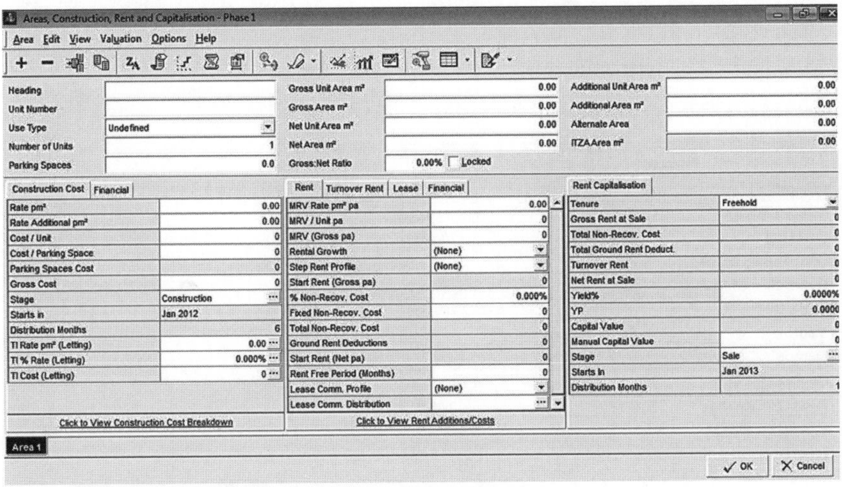

Figure 12.38 The capitalised rent area screen

else estimated using the gross to net ratio function. Note the program allows the modelling of multiple units of the same size/type. Other areas that can be defined here are parking spaces and additional and alternate areas. Here I am using the additional areas for the landscaping and car park areas which will be costed later in the appraisal (Figure 12.39).

The mid-left area of the screen is where the main construction cost calculation and modelling takes place. The construction frame (Figure 12.40) takes the measurement data entered into the program, using the gross area to calculate the costs plus any entry for parking spaces etc. The cost data can be entered in one of four ways; either by a rate per unit of measurement (in this case m^2), a cost per unit, the overall unit cost or else by way of a detailed construction cost breakdown if a detailed estimate is available.

Note that the TI boxes in the bottom part of this frame relate to tenants improvements, works that can be scheduled to meet the requirements of an individual tenant and where a separate cost estimate is required.

Construction cost information can be surprisingly hard to obtain. Most up-to-date cost information is collected and sold by commercial organisations such as the BCIS, the Building Cost Information Service owned by the RICS, and price book publishers such as Laxtons and Wessex. The situation is made more complex by the extremes in range in costs displayed in construction; similarly specified buildings would have very different construction costs depending on

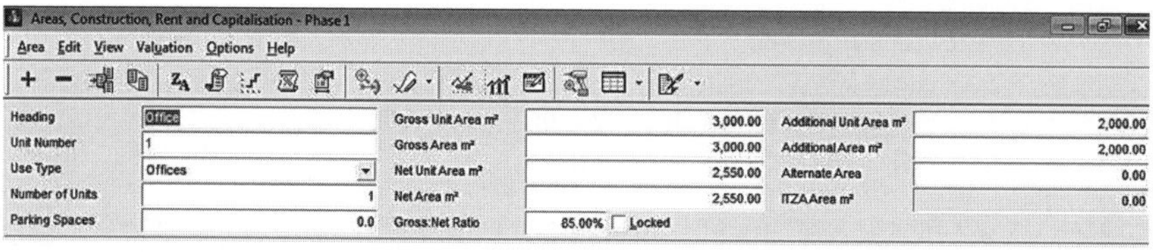

Figure 12.39 Use type and areas

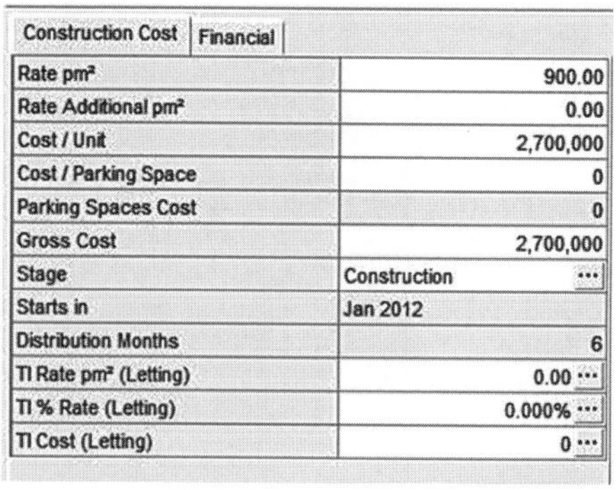

Figure 12.40 Construction calculation frame

whether they were located in central London or in a greenfield, edge of town location whilst the cost of constructing an office can easily range from £600 to £4000 per m² dependent upon type, specification and location. Anyone conducting a development appraisal would be well advised to consult a construction cost specialist and to look at the specification and location very carefully.

The timing and distribution of the construction costs can be viewed by pressing the shortcut button on the stage row (Figure 12.41). The timing and distribution can be amended by checking allow custom distribution. The data can then be manipulated using the three edit boxes available or else by clicking into the data distribution editor which opens up a mini-spreadsheet. A fifth amendment option is to do the revisions in the cash flow tab of the program. It should be noted that the amendments made to any individual item do not affect the global timing assumptions. It is thus easily possible to adjust the timing of just one building amongst a group of buildings in isolation.

In this case I am content with the distribution that the program has given me and I will make no adjustments.

Note that there is a second frame behind the calculation area which deals with finance. This allows the allocation of a different interest set and the individual (rather than global) treatment of VAT. There is no absolute need to adjust these values and I will leave them in the default setting for this appraisal.

The lower centre part of the area screen is where the income screen for the developed property is calculated (Figure 12.42). There are three possible modes of data entry, a rate per unit area, a Market Rental Value (MRV) for

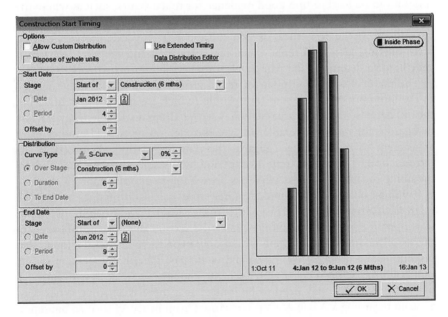

Figure 12.41 Construction cost distribution

Rent	Turnover Rent	Lease	Financial	
MRV Rate pm² pa			180.00	▲
MRV / Unit pa			459,000	
MRV (Gross pa)			459,000	
Rental Growth	(None)		▼	
Step Rent Profile	(None)		▼	
Start Rent (Gross pa)			459,000	
% Non-Recov. Cost			0.000%	
Fixed Non-Recov. Cost			0	
Total Non-Recov. Cost			0	
Ground Rent Deductions			0	
Start Rent (Net pa)			459,000	
Rent Free Period (Months)			6	
Lease Comm. Profile	(None)		▼	
Lease Comm. Distribution			...	▼
Click to View Rent Additions/Costs				

Figure 12.42 Rental calculation frame

each unit and an overall (Gross) MRV (which is the same as the previous when there is only one unit).

Rental value is determined by what tenants will pay to lease property of a similar standard and in similar locations in the open market. In active markets this can be relatively easy to determine, other than the fact that the development will be completed at some time in the future during which time the economy and local market conditions may have changed radically. In other places and times it can be hard to find good evidence for rental values, such as where the building being developed is unusual for the area or when market conditions are difficult. In the downturns of 1990–93 and 2008 onwards, property owners were offering such high rental incentives (rent frees and the like) that it was very hard to determine the true level of rental value in many markets.

The frame allows adjustment for rental growth (which have to be defined in the assumptions tab), a step rent profile (which has also to be defined but this is done within the area screen via the drop down menus) and any deductions for non-recoverable parts of the income stream. The other main adjustment to be made here regards the rent free period. As noted in the introduction, a 6 month rent free period was assumed to be required by incoming tenants and this is what has been entered into the appropriate box.

It will be observed that there are three other calculation frames behind the main calculation frame. The first allows turnover rents, most commonly found in shopping centres, to be modelled. In the second the lease assumptions in the template are laid out. These would only need to be altered if a long-term holding of the property was envisaged or if some special arrangement (e.g. a rent holiday) had been made with a tenant. The final frame deals with financial matters again. This may need to be looked at if there were some issues with a tenant regarding VAT (e.g. a VAT exempt tenant in a VAT elected building). In the main, however, no entry is usually required in any of these frames.

The centre lower right part of the screen is given over the calculation of income frame (Figure 12.43). As the name implies this is where the data from the income stream frame is capitalised to determine the end value of this part of the development output. By default, Developer assumes that the interest is freehold, but the program can be switched via the drop down menu to a leasehold interest. This creates a frame behind the front screen that allows the entry of the head lease details such as rent, review pattern and duration. In this case we are dealing with a freehold property, however.

There are two modes of calculation of the capital value of the freehold interest, either by entering a yield into the appropriate box or else by entering a manual value. This might be required where a presale agreement had been entered into.

Yield, like rental value, is market determined. It is essentially the annual return that the investment owner will accept given the quality of income stream (determined by a complex combination of the quality of the tenant, the length and quality of the lease, the quality of the building and the attractiveness of the location amongst other things) and the returns obtainable on rival investments, both property and others. Office yields have as high a range as office rents and office costs. Careful research is required to place the completed investment in its correct standing in the market.

As with all elements of Developer, it is possible to adjust the timing and distribution of the projected sale receipts by clicking the three dots next to the 'stage' row and bringing up the distribution screen (Figure 12.44). The same options for altering the distribution and timing as were available with the construction section are available here and, indeed, with all other elements of the appraisal.

When the area screen has been completed (Figure 12.45), then the green 'OK' button is clicked to save the assumptions made. This then returns us to the definitions tab (Figure 12.46).

It will be noted that the key values from the area screen have been imported in summary to the definitions tab, namely the construction costs and completed

Rent Capitalisation		
Tenure	Freehold	▼
Gross Rent at Sale		459,000
Total Non-Recov. Cost		0
Total Ground Rent Deduct.		0
Turnover Rent		0
Net Rent at Sale		459,000
Yield%		6.0000%
YP		16.6667
Capital Value		7,650,000
Manual Capital Value		0
Stage	Sale	...
Starts In	Jan 2013	
Distribution Months		1

Figure 12.43 The capitalisation of income frame

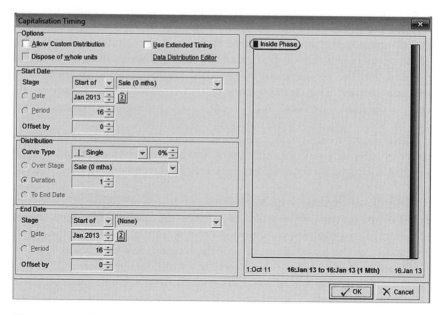

Figure 12.44 The data distribution screen for the disposal of the freehold interest

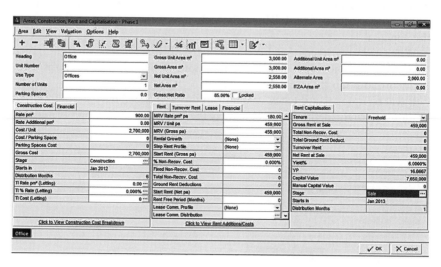

Figure 12.45 The completed area tab for our development

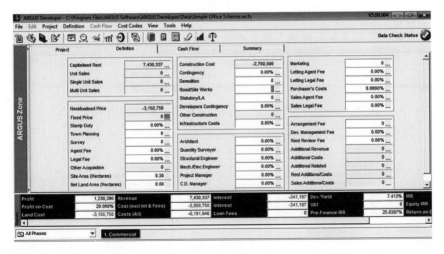

Figure 12.46 Definitions tab after the area screen has been completed

value. Both boxes have been greyed out to indicate that they have been calculated outside of the definitions tab itself. No direct edit of these values is possible from here. Any changes have to be made by clicking back into the area tab or else adjusting the values and distribution directly in the cash flow.

The appraisal is completed by entering the required values in the definitions tab. This can be done either by entering figures directly into the appropriate open box on this screen and letting the program distribute the values according to the template (this would be done if a quick appraisal is required or if there are limited details of the scheme available as in an early stage feasibility study) or else drilling down into each box and entering the details of the expenditure or receipts.

In this case, I have used the quick appraisal approach with the exception of the road/site works box where I have modelled the hard and soft landscaping by reference to the additional area I defined in the area tab. The default setting in the template is a single lump sum item tied to the beginning of the construction stage. Here I have over-ridden the default, turning the item into a cost related to the alternative area. I have also adjusted the timing so that the expenditure occurs towards the end of the construction stage, as would occur in practice (indeed, I have allowed the landscaping to run on into the letting up phase – see Figure 12.47). In this I can illustrate the flexibility and power of Developer. You can use the program in its default mode and it will produce a quick and reasonably accurate appraisal, or, if you have more information or there are some special assumptions, you can take control of any or all of the items making up the development and produce a tailored, more accurate model of the actual development project.

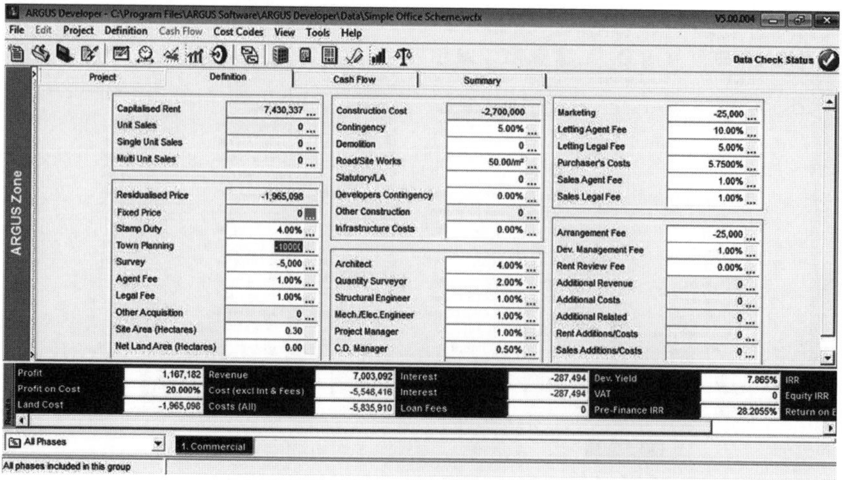

Figure 12.47 Detail of the hard landscaping assumptions override

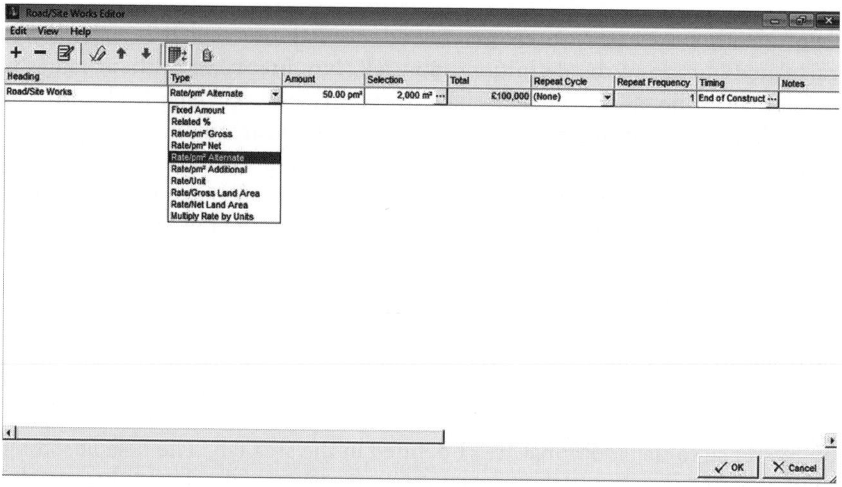

Figure 12.48 The completed definitions tab

It will be noted that the definitions tab (Figure 12.48) is itself divided into functional areas for ease of calculation. The area to the bottom left of the screen is generally connected with the site, its acquisition and any investigations required pre and post purchase. There is also a section dealing with the costs of getting planning consent.

The upper part of the middle part of the definitions tab is where the main construction cost is reported and where any additional construction cost items are located. There are two contingency cells where allowances for uncertainty in construction and other aspects of the development can be allowed for. There is also a section that deals with s106 agreements (the payments or work required by the local authority as part of the planning consent granted for the development) that is hidden in the 'additional construction costs' cell. This often has to be enabled in the software (under file/administration/system configuration/country).

The lower centre part of the definitions tab is where the cost of the construction professional team is calculated. The default setting is that these are related to the construction costs, both those defined in the area screen and any additional costs. These settings can be overridden by drilling down into the calculation screen behind the cell (Figure 12.49). This might be required where the professional team are working to fixed fees or where some special arrangement is required.

The top left part of the definitions tab is where the disposal of the development is modelled. This includes cells for marketing, letting and sale fees and also purchaser's costs. The latter is to account for the adjustment to the price paid that someone buying the investment would make to allow for the costs they incur in the acquisition including their own legal and surveyors costs and the taxes (stamp duty) they will be charged. To give a simple example, if an investment produces a rent of £100,000 and the investor requires a 10 per cent return you would expect them to pay £1,000,000 for it. If no allowance is made for their acquisition costs (agents fees of 1 per cent, legal fees of 1 per cent and stamp duty of 4 per cent) the total paid would be £1,060,000 which would give them a yield of only 9.43 per cent. The price paid has to be netted down to total £1,000,000 after acquisition costs to give

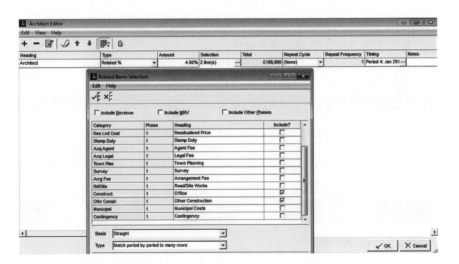

Figure 12.49 Overriding the related items setup

the true yield, and this is what the program does automatically when an entry is made here.

The lower right part of the definition tab is mainly a sweep up area where any other costs or revenue streams not accounted for elsewhere can be modelled. It is worth noting once again that Developer's flexibility allows for most of the cells on the definition tab to be renamed or tailored to meet the specific requirements of the scheme being appraised.

The data entered into the definition tab and any associated timing assumptions are automatically placed in the cash flow (Figure 12.50) according to the template. Each data entry creates its own line in the cash flow and it is to the cash flow that we would normally go to once all the data has been input into the distribution tab.

There are two cash flows in Developer though they are both interlinked, the project cash flow and the finance cash flow. The project cash flow shows each of the individual income and expenditure items in the assumed timeframe and distribution either according to that laid down by the template or according to the individual distribution determined by the modeller. These cash flows are, in the main, open to be altered; as noted in the introduction the cash flow in the program is the primary calculation tool, it is an output but it is also a tool for fine tuning the appraisal model (Figure 12.51).

This can be illustrated using the marketing line. The standard template puts the marketing expenditure into the beginning of the letting period, which is a not unreasonable approximation for an initial feasibility study. It would be more accurate, however, to spread the expenditure over part of the construction phase as it is almost certain that the marketing campaign would start far earlier than this to ensure that the income stream would start as soon as possible (there

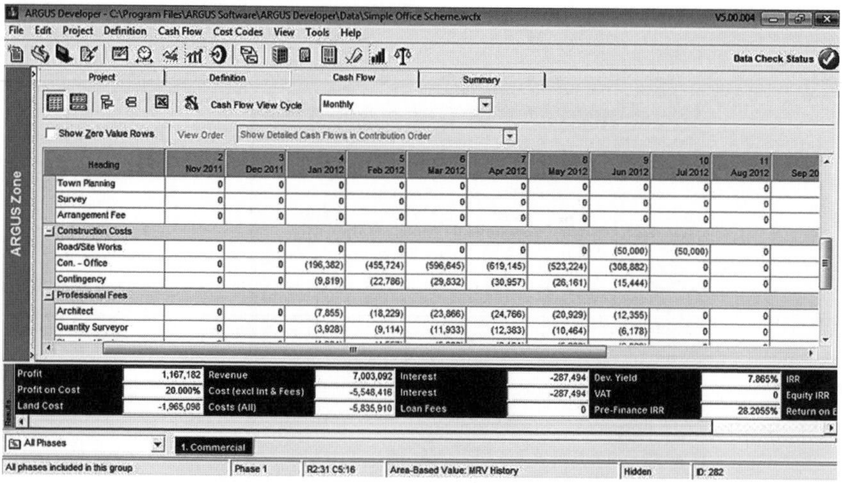

Figure 12.50 The cash flow tab

Heading	1 Oct 2011	2 Nov 2011	3 Dec 2011	4 Jan 2012	5 Feb 2012	6 Mar 2012	7 Apr 2012	8 May 2012	9 Jun 2012	10 Jul 2012	Aug 20
Architect	0	0	0	(7,855)	(18,229)	(23,866)	(24,766)	(20,929)	(12,355)	0	
Quantity Surveyor	0	0	0	(3,928)	(9,114)	(11,933)	(12,383)	(10,464)	(6,178)	0	
Structural Engineer	0	0	0	(1,964)	(4,557)	(5,966)	(6,191)	(5,232)	(3,089)	0	
Mech./Elec.Engineer	0	0	0	(1,964)	(4,557)	(5,966)	(6,191)	(5,232)	(3,089)	0	
Project Manager	0	0	0	(1,964)	(4,557)	(5,966)	(6,191)	(5,232)	(3,089)	0	
C.D. Manager	0	0	0	(982)	(2,279)	(2,983)	(3,096)	(2,616)	(1,544)	0	
− Marketing/Letting											
Marketing	0	0	0	0	0	0	0	0	0	(25,000)	
Letting Agent Fee	0	0	0	0	0	0	0	0	0	0	
Letting Legal Fee	0	0	0	0	0	0	0	0	0	0	

Figure 12.51 Adjusting the development model via the cash flow

is a strong imperative for the developer/trader of a speculative development to achieve a letting as the investment is not saleable unless it is 100 per cent let. Without the sale a developer/trader cannot recoup their development costs and realise their profits). The distribution of this line can be altered in the definition tab but can also be adjusted via the cash flow. This can be done in two ways: the first by clicking into the line and selecting data distribution from the drop down menu (Figure 12.52).

This opens up the data distribution screen pertinent to this line and the process for adjustment is as discussed in earlier sections. The alternative way of adjusting

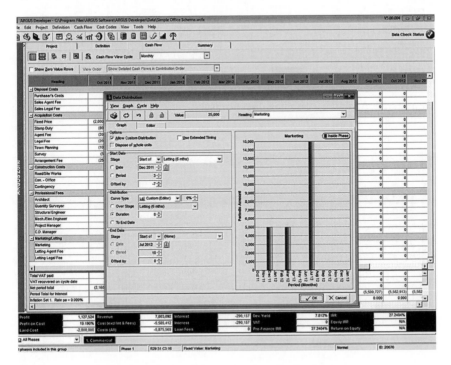

Figure 12.52 Adjusting a cash flow line via the data distribution

the distribution is to type directly into the cash flow (Figure 12.53). This is only possible for lines which are not dependent ones, i.e. not calculated by reference to some other line, as is the case with marketing expenditure.

This allows a direct adjustment to the sums originally modelled that may have come when more information on the scheme is received or new estimates – the sum assessed may well go up or down – but it can create problems where the item being adjusted is complex. This remains a very useful modelling tool, however.

The second cash flow is the finance cash flow (Figure 12.54). This shows the results of the cash flow 'below the line', i.e. the result of the cash flow calculations including interest calculations and VAT (if allowed for in the calculation). The cash flow items in this section are not adjustable.

Once the cash flow has been reviewed and adjusted, if required, then it is normal to go to the appraisal summary, which, as noted, is the standard output of the appraisal, summarising the project in a simple and clear manner. This can be either printed off directly or else incorporated into the reporting module.

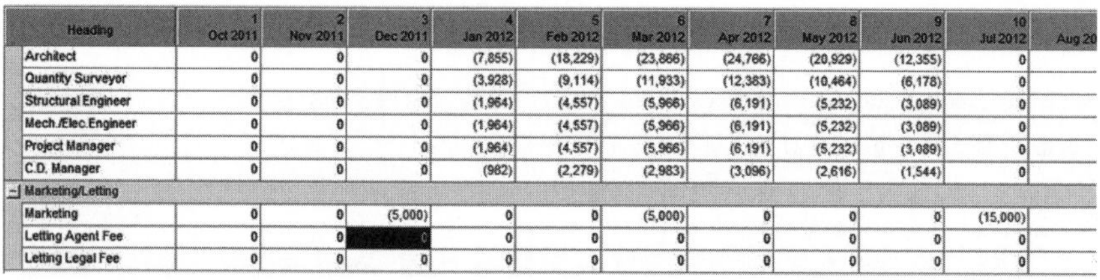

Heading	1 Oct 2011	2 Nov 2011	3 Dec 2011	4 Jan 2012	5 Feb 2012	6 Mar 2012	7 Apr 2012	8 May 2012	9 Jun 2012	10 Jul 2012	Aug 20
Architect	0	0	0	(7,855)	(18,229)	(23,866)	(24,766)	(20,929)	(12,355)	0	
Quantity Surveyor	0	0	0	(3,928)	(9,114)	(11,933)	(12,383)	(10,464)	(6,178)	0	
Structural Engineer	0	0	0	(1,964)	(4,557)	(5,966)	(6,191)	(5,232)	(3,089)	0	
Mech./Elec.Engineer	0	0	0	(1,964)	(4,557)	(5,966)	(6,191)	(5,232)	(3,089)	0	
Project Manager	0	0	0	(1,964)	(4,557)	(5,966)	(6,191)	(5,232)	(3,089)	0	
C.D. Manager	0	0	0	(982)	(2,279)	(2,983)	(3,096)	(2,616)	(1,544)	0	
Marketing/Letting											
Marketing	0	0	(5,000)	0	0	(5,000)	0	0	0	(15,000)	
Letting Agent Fee	0	0	0	0	0	0	0	0	0	0	
Letting Legal Fee	0	0	0	0	0	0	0	0	0	0	

Figure 12.53 Adjusting a cash flow line manually

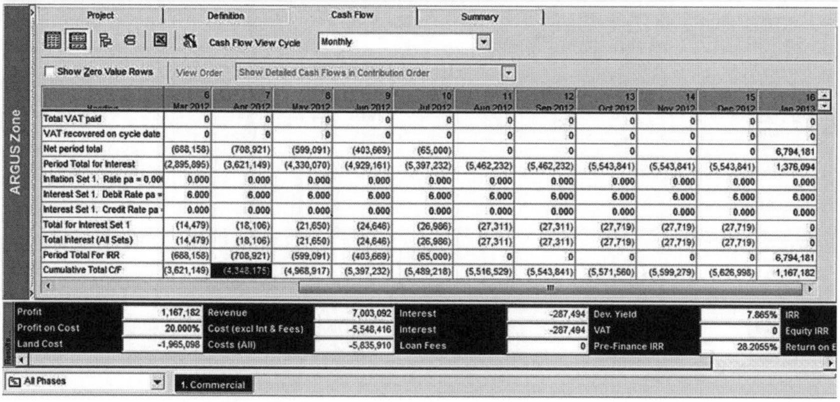

Heading	6 Mar 2012	7 Apr 2012	8 May 2012	9 Jun 2012	10 Jul 2012	11 Aug 2012	12 Sep 2012	13 Oct 2012	14 Nov 2012	15 Dec 2012	16 Jan 2013
Total VAT paid	0	0	0	0	0	0	0	0	0	0	0
VAT recovered on cycle date	0	0	0	0	0	0	0	0	0	0	0
Net period total	(688,158)	(708,921)	(599,091)	(403,669)	(65,000)	0	0	0	0	0	6,794,181
Period Total for Interest	(2,895,895)	(3,621,149)	(4,330,070)	(4,929,161)	(5,397,232)	(5,462,232)	(5,462,232)	(5,543,841)	(5,543,841)	(5,543,841)	1,376,094
Inflation Set 1. Rate pa = 0.00	0.000	0.000	0.000	0.000	0.000	0.000	0.000	0.000	0.000	0.000	0.000
Interest Set 1. Debit Rate pa =	6.000	6.000	6.000	6.000	6.000	6.000	6.000	6.000	6.000	6.000	6.000
Interest Set 1. Credit Rate pa	0.000	0.000	0.000	0.000	0.000	0.000	0.000	0.000	0.000	0.000	0.000
Total for Interest Set 1	(14,479)	(18,106)	(21,650)	(24,646)	(26,986)	(27,311)	(27,311)	(27,719)	(27,719)	(27,719)	0
Total Interest (All Sets)	(14,479)	(18,106)	(21,650)	(24,646)	(26,986)	(27,311)	(27,311)	(27,719)	(27,719)	(27,719)	0
Period Total For IRR	(688,158)	(708,921)	(599,091)	(403,669)	(65,000)	0	0	0	0	0	6,794,181
Cumulative Total C/F	(3,621,149)	(4,348,175)	(4,968,917)	(5,397,232)	(5,489,218)	(5,516,529)	(5,543,841)	(5,571,560)	(5,599,279)	(5,626,998)	1,167,182

Profit	1,167,182	Revenue	7,003,092	Interest	-287,494	Dev. Yield	7.865% IRR
Profit on Cost	20.000%	Cost (excl Int & Fees)	-5,548,416	Interest	-287,494	VAT	0 Equity IRR
Land Cost	-1,965,098	Costs (All)	-5,835,910	Loan Fees	0	Pre-Finance IRR	28.2055% Return on E

Figure 12.54 The finance cash flow

The appraisal summary for our development is displayed below.

APPRAISAL SUMMARY
Single Building Office Development

Summary Appraisal for Part 1 Commercial

REVENUE

Rental Area Summary				*Initial*	*Net Rent*
	Units	*m²*	*Rate m²*	*MRV/Unit*	*at Sale*
Office	1	2,550.00	£180.00	£459,000	459,000

Investment Valuation
   Office

Market Rent 459,000	YP @	6.0000%	16.6667		
(0yrs 6mths	PV 0y	6.0000%	0.9713	7,430,337	
Rent Free)	6m @				

GROSS DEVELOPMENT VALUE			7,430,337
Purchaser's	5.75%	(427,244)	
Costs			

NET DEVELOPMENT VALUE   7,003,092

NET REALISATION   7,003,092

OUTLAY

ACQUISITION COSTS

Residualised Price (0.30 Ha			1,965,098
£6,550,327.12 pHect)			
Stamp Duty	4.00%	78,604	
Agent Fee	1.00%	19,651	
Legal Fee	1.00%	19,651	
Town Planning		10,000	
Survey		5,000	
			2,098,004

CONSTRUCTION COSTS

Base Construction 3,000.00 m²		2,700,000
@ £900.00 pm²		
Contingency	135,000	
Road/Site Works	100,000	
		2,935,000

PROFESSIONAL FEES

Architect	4.00%	108,000
Quantity Surveyor	2.00%	54,000
Structural Engineer	1.00%	27,000
Mech./Elec.Engineer	1.00%	27,000

Project Manager	1.00%	27,000	
C.D. Manager	0.50%	13,500	
			256,500

MARKETING & LETTING

Marketing		25,000	
Letting Agent Fee	10.00%	45,900	
Letting Legal Fee	5.00%	22,950	
			93,850

DISPOSAL FEES

Sales Agent Fee	1.00%	70,031	
Sales Legal Fee	1.00%	70,031	
			140,062

Total Additional Costs   25,000

FINANCE

Debit Rate 6.000% Credit Rate 0.000% (Nominal)

Land	84,418	
Construction	38,310	
Letting Void	164,766	
Total Finance Cost		287,494

TOTAL COSTS   5,835,910

PROFIT

                                                     1,167,182

Performance Measures

Profit on Cost%	20.00%
Profit on GDV%	15.71%
Profit on NDV%	16.67%
Development Yield% (on Rent)	7.87%
Equivalent Yield% (Nominal)	6.00%
Equivalent Yield% (True)	6.23%
IRR	28.21%
Rent Cover	2 yrs 7 mths
Profit Erosion (finance rate 6.000%)	3 yrs 1 mth

It is somewhat difficult to see the goal of our calculation, namely the residualised land value but it is there under the section headed Acquisition Costs at a value of £1,965,098, this being the maximum sum that should be paid for the land to achieve the target figure of 20 per cent profit on cost.

We will assume that this is the end of the appraisal and no other calculation is required.

Argus Developer case study two – profitability calculation

As noted earlier, development appraisals are most commonly done for one of two main purposes; firstly to calculate the amount the developer can bid for the site and still meet their target profit figure (assuming that their assumptions are valid). To do this the appraiser has to fix the profit percentage, it becomes the known in the equation. The alternative calculation is to calculate the profitability of a scheme. To do this the land value or rather price has to be known, the unknown variable, the residual sum of the calculation is the profit figure.

With Developer this calculation is simple. Essentially the data entry entered is the same, it just requires the set-up of the programme to be slightly different.

Let us assume that the developer can purchase the land for the development at £2,000,000.

The first change is to switch the programme into Fixed Land Value mode in the assumptions for calculation (Figure 12.55). As noted earlier, this is the default setting for the program.

The second change required is to enter the land purchase price into the now open box in the definition tab (Figure 12.56).

All other entries in the programme are as per the previous calculation discussed above.

The summary for this appraisal is as follows.

APPRAISAL SUMMARY
Single Building Office Development

Summary Appraisal for Part 1 Commercial

REVENUE

Rental Area Summary				Initial	Net Rent
	Units	m^2	Rate m^2	MRV/Unit	at Sale
Office	1	2,550.00	£180.00	£459,000	459,000

Figure 12.55 Assumptions for calculation set-up for profit calculation

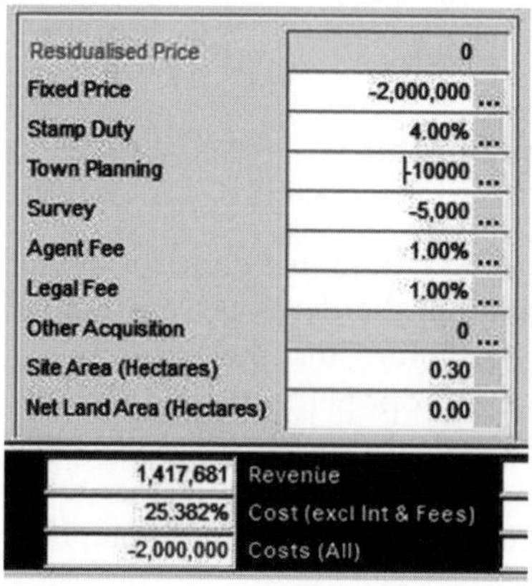

Figure 12.56 Land value entered into definition tab

Investment Valuation
  Office
    Market Rent 459,000   YP @   6.0000%   16.6667
    (0yrs 6mths   PV 0y 6m @   6.0000%   0.9713   7,430,337
      Rent Free)

GROSS DEVELOPMENT VALUE   7,430,337
    Purchaser's Costs   5.75%   (427,244)

NET DEVELOPMENT VALUE   7,003,092

NET REALISATION   7,003,092

OUTLAY

ACQUISITION COSTS
    Fixed Price (0.30 Ha   2,000,000
      £6,666,666.67 pHect)
    Stamp Duty   4.00%   80,000
    Agent Fee   1.00%   20,000
    Legal Fee   1.00%   20,000
    Town Planning   10,000
    Survey   5,000
                                              2,135,000

CONSTRUCTION COSTS
    Base Construction 3,000.00 m²   2,700,000
      @ £900.00 pm²
    Contingency   135,000
    Road/Site Works   100,000
                                 2,935,000

PROFESSIONAL FEES

Architect	4.00%	108,000
Quantity Surveyor	2.00%	54,000
Structural Engineer	1.00%	27,000
Mech./Elec. Engineer	1.00%	27,000
Project Manager	1.00%	27,000
C.D. Manager	0.50%	13,500
		256,500

MARKETING & LETTING

Marketing		25,000
Letting Agent Fee	10.00%	45,900
Letting Legal Fee	5.00%	22,950
		93,850

DISPOSAL FEES

Sales Agent Fee	1.00%	70,031
Sales Legal Fee	1.00%	70,031
		140,062

Total Additional Costs	25,000

FINANCE

Debit Rate 6.000% Credit Rate 0.000% (Nominal)

Land	85,918	
Construction	38,310	
Letting Void	165,930	
Total Finance Cost		290,157

TOTAL COSTS   5,875,569

PROFIT

                                     1,127,524

Performance Measures	
Profit on Cost%	19.19%
Profit on GDV%	15.17%
Profit on NDV%	16.10%
Development Yield% (on Rent)	7.81%
Equivalent Yield% (Nominal)	6.00%
Equivalent Yield% (True)	6.23%
IRR	27.24%
Rent Cover	2 yrs 5 mths
Profit Erosion (finance rate 6.000%)	2 yrs 11 mths

The performance measures that conclude the appraisal summary are very important outcomes for the developer. They represent the key performance benchmarks that developers and financiers use to assess whether a project is viable. The normal benchmark for a commercial scheme is 20 per cent so this scheme is rather marginal at this land cost.

More complex projects

Introduction

Argus Developer is ideally suited to complexity, indeed this is one of the program's key advantages over self-constructed Excel models which often struggle to adapt to increasing complexity and, in particular, scheme alterations.

This section looks at examples of more complex (and thus often more realistic projects for the real world where complexity is the norm rather than the exception – each development is often unique) projects that are often encountered in practice, illustrating in each how Developer can solve the problems in appraisal faced.

Because the setup, data entry and output are essentially the same as covered in our simple commercial project this section will concentrate on the specific items of variance required for the topic.

Argus Developer case study three – multiple buildings/ types

Often a development involves the construction of more than one type of building at a time. This is frequently the case in larger sites and also in mixed use developments. These often mix commercial and residential uses together. Residential development appraisal has its own section in this book because of its special characteristics; however it seems appropriate to include a residential element in one of the examples as this will be a common component of a multiple type scenario in practice.

The first example we will look at however will be a mixed commercial scheme that has two 1000m² and three 750m² office buildings, plus 5000m² of retail warehousing in 4 buildings, two of 1500m² and two of 1000m².

This development will also form the basis of the multiple timings/single phase development below but for this initial development we will assume that they will occur in the same timeframe and not attempt to fine-tune the timings.

This initial appraisal will be run as a land residual (Figure 12.57).

As noted above, the basic procedure for setting up the project is identical for that which was followed for the simple scheme, however for the record the interest rate used here is a flat rate of 6 per cent in a single interest set and the target profit figure for the land residual is 20 per cent profit on cost.

No timescales were defined in the project brief but here I have assumed an immediate purchase of the site, six months of planning and pre-construction

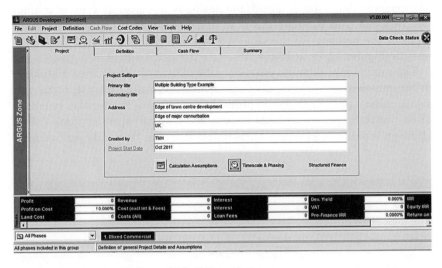

Figure 12.57 Project tab for multi-building scheme

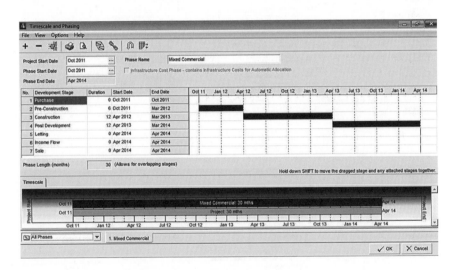

Figure 12.58 Timescales for the multi-building project

followed by a 12 month build period and a 12 month letting up period before the development is sold (Figure 12.58).

Once these parameters have been entered into the program, we can move to the definition tab. The way that we can deal with the problem of multiple buildings is very simple; we must create multiple tabs in the commercial area drill down screen.

We have a total of nine buildings, five offices and four retail warehouses. We could create a separate area record for each building (and there are great advantages in doing this as regards the flexibility this gives in terms of timing) but for speed (and assuming this is an initial feasibility study) we will group together the buildings of the same size and use giving us four separate area records (Figures 12.59 to 12.62).

This subdivision, although preventing the individual timing of each building in the sub-class (see below), does allow minor variations in cost, rental and capital value items which would probably occur in practice (smaller buildings often cost more per m^2 to build because the ratio of expensive items (walls, cladding, windows and doors) to floor area is greater than in larger buildings

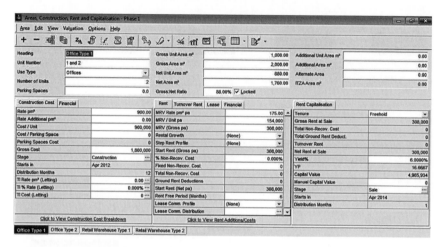

Figure 12.59 Office Type A area record

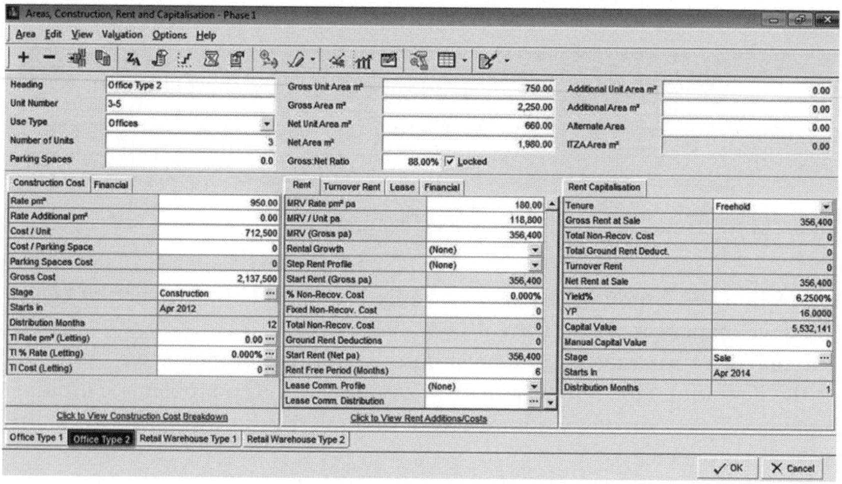

Figure 12.60 Office Type B area record

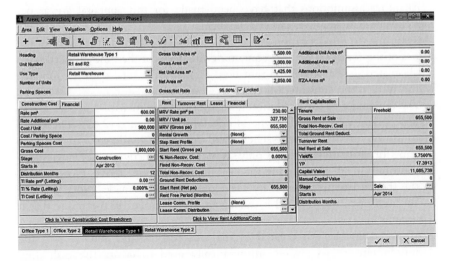

Figure 12.61 Retail Warehouse Type 1 area record

Figure 12.62 Retail Warehouse Type 2 area record

and working space is reduced leading to inefficiencies in on-site activities. Conversely there is sometimes a discount for size in rents paid per m² for larger buildings (or a premium for smaller, the effect is the same). Larger buildings often attract bigger companies, usually with greater financial stability than those who occupy smaller premises therefore the yield is often bid down for larger buildings which causes the values to rise.). These considerations have been factored into the area records.

All other assumptions that have been made for this appraisal are as per Developer's template (see earlier reference).

Returning to the definition tab, it can be observed that the total sum of the values of the four area records containing the detail of the nine properties are amalgamated in the capitalised rent box. Similarly the total construction cost is displayed in the construction cost box.

The remainder of the definition tab has been filled in with what are felt to be appropriate values for this type of development (Figure 12.63). These will of course vary from scheme to scheme and the reader should view these values as being for illustrative purposes only. A more detailed coverage of entries into the distribution tab are found in the introductory section, above. No variation to the standard template used by Developer has been made. Note that the residualised price for the land is calculated by Developer in the mode that it is running in.

All the data that was created in the area records and entered into the definition tab are, as always, automatically used to create the cash flow (Figure 12.64).

As can be seen from the extract from the cash flow, the template builds all the buildings together side by side. This probably would not occur in practice but is sufficiently accurate for an initial feasibility study to let it ride. Similarly, all of the buildings are assumed to be sold and let at the same time, namely at the end of the letting up period, which is also rather unrealistic but which would also be a normal assumption for an initial, rough appraisal done to test the basic feasibility of the scheme. The question of timings will be addressed below.

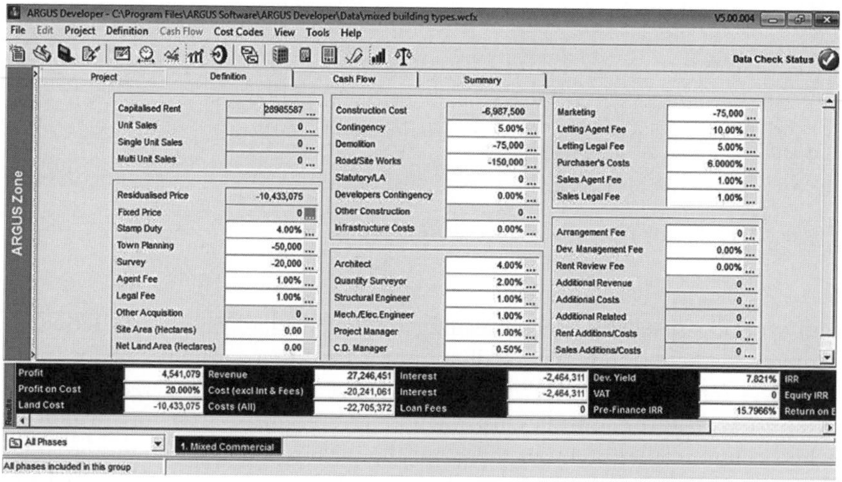

Figure 12.63 The completed definition tab for the multi-building scheme

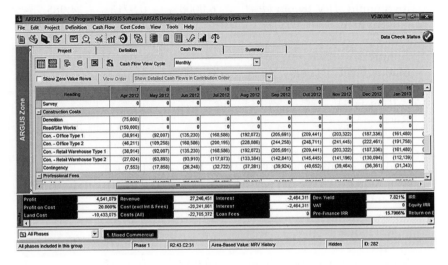

Figure 12.64 Cash flow for the multi-building scheme

The appraisal summary for this scheme is laid out below.

APPRAISAL SUMMARY
Multiple Building Type Example

Summary Appraisal for Part 1 Mixed Commercial

REVENUE

Rental Area Summary

	Units	m^2	Rate m^2	Initial MRV/Unit	Net Rent at Sale
Office Type 1	2	1,760.00	£175.00	£154,000	308,000
Office Type 2	3	1,980.00	£180.00	£118,800	356,400
Retail Warehouse Type 1	2	2,850.00	£230.00	£327,750	655,500
Retail Warehouse Type 2	2	1,900.00	£240.00	£228,000	456,000
Totals	9	8,490.00			1,775,900

Investment Valuation
  Office
    Type 1
    Market Rent 308,000 YP @   6.0000%   16.6667

(0yrs 6mths PV 0y 6m @		6.0000%	0.9713	4,985,934
Rent Free)				

Office
  Type 2

Market Rent 356,400 YP @		6.2500%	16.0000	
(0yrs 6mths PV 0y 6m @		6.2500%	0.9701	5,532,141
Rent Free)				

Retail
  Warehouse
  Type 1

Market Rent 655,500 YP @		5.7500%	17.3913	
(0yrs 6mths PV 0y 6m @		5.7500%	0.9724	11,085,739
Rent Free)				

Retail
  Warehouse
  Type 2

Market Rent 456,000 YP @		6.0000%	16.6667	
(0yrs 6mths PV 0y 6m @		6.0000%	0.9713	7,381,773
Rent Free)				
				28,985,587

GROSS DEVELOPMENT VALUE			28,985,587
Purchaser's Costs	6.00%	(1,739,135)	

NET DEVELOPMENT VALUE   27,246,451

NET REALISATION   27,246,451

OUTLAY

ACQUISITION COSTS

Residualised Price		10,433,075	
Stamp Duty	4.00%	417,323	
Agent Fee	1.00%	104,331	
Legal Fee	1.00%	104,331	
Town Planning		50,000	
Survey		20,000	
			11,129,059

CONSTRUCTION COSTS

Base Construction 9,250.00 m²	6,987,500	
Contingency	349,375	
Demolition	75,000	
Road/Site Works	150,000	
		7,561,875

PROFESSIONAL FEES

Architect	4.00%	279,500
Quantity Surveyor	2.00%	139,750

Structural Engineer	1.00%	69,875	
Mech./Elec. Engineer	1.00%	69,875	
Project Manager	1.00%	69,875	
C.D. Manager	0.50%	34,938	
			663,813

MARKETING & LETTING

Marketing		75,000	
Letting Agent Fee	10.00%	177,590	
Letting Legal Fee	5.00%	88,795	
			341,385

DISPOSAL FEES

Sales Agent Fee	1.00%	272,465	
Sales Legal Fee	1.00%	272,465	
			544,929

FINANCE

Debit Rate 6.000% Credit Rate 0.000% (Nominal)

Land	1,712,206	
Construction	752,105	
Total Finance Cost		2,464,311

TOTAL COSTS   22,705,372

PROFIT

                                   4,541,079

Performance Measures

Profit on Cost%	20.00%	
Profit on GDV%	15.67%	
Profit on NDV%	16.67%	
Development Yield% (on Rent)	7.82%	
Equivalent Yield% (Nominal)	5.95%	
Equivalent Yield% (True)		6.18%
IRR	15.80%	
Rent Cover	2 yrs 7 mths	
Profit Erosion (finance rate 6.000%)	3 yrs 1 mth	

It can be seen that Developer copes with multiple building types very easily. The program is highly flexible and well thought out which is not surprising after over 20 years of development and the feedback from the many thousands of Developer users. It also allows easy alteration of these details without issue.

Argus Developer case study four – mixed-use buildings

Still under the heading of multiple buildings, let us examine the case of a mixed use building, i.e. different uses within the same building. As noted in the introduction to this section the most common type of mixed use building involves residential property, particularly what is more common in city centres, residential apartments over a commercial ground floor.

The reason why I wanted to specifically cover this type of development here, even though there is a separate section for residential development (and there will be some inevitable repetition of some area for which I apologise), is that there are very specific timing issues which have to be addressed to produce a more accurate appraisal in mixing commercial and residential properties in the same appraisal (Figure 12.65).

The program is basically set up in the same way as for the previous developments, so I will not step through this process for this scheme, other than to say the interest rate assumed is 6 per cent and the total development period is 19 months, which includes a 3 month pre-construction period and a 9 month construction phase. The remaining details of timing will be discussed below. Also you will note that rather than step through the build up for the appraisal we are analysing the make-up of a completed appraisal.

Reference to the completed development tab (Figure 12.66) illustrates that both the capitalised rent, used primarily for commercial property, and the unit sales, used mainly for residential units sold on, areas have been used in this appraisal.

Drilling down into the area screen for the capitalised rent, you will observe that I have taken the option of using the Zone A calculator to determine the rental value. Standard shop units in UK practice are frequently 'zoned', that is

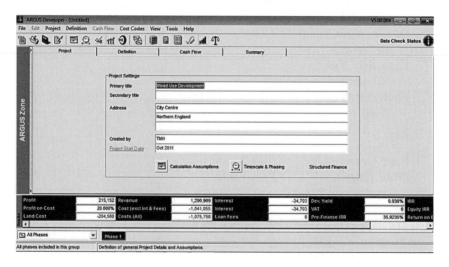

Figure 12.65 Project tab for the mixed-use development example

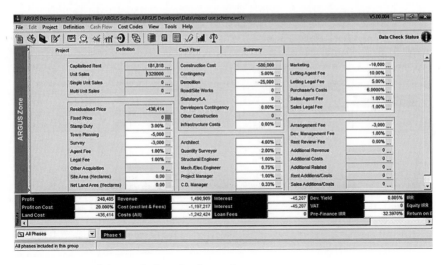

Figure 12.66 Definition tab for mixed use scheme

the area is calculated under the assumption that the most valuable part of the shop is that at the front and, beyond a certain point, that value reduces. Practice in most of the UK is to assume zone depths of 6 metres, with the first zone valued at 100 per cent, the next with a 50 per cent reduction, and the third at 25 per cent of the value of zone A. The remainder of the trading floor beyond Zone C, storage and upper and lower floor trading have a separate treatment.

Developer has a simple but useful Zone A calculator (accessed from the area screen via the ZA button) which both calculates the net area of the scheme and also the rental value according to the user defined ITZA (In Terms of Zone A) rent, both of which are pasted into the relevant parts of the area screen (Figures 12.67 and 12.68).

It is necessary to fill in the gross (built) area so that the construction cost can be calculated. All other elements are as we have already covered.

The unit sales area is the basic data entry point for residential development (Figure 12.69). It used to be the only area where residential developments were modelled until the more sophisticated area sheets were added in version 4 of Developer (these will be covered in the specific section on residential development). The unit sales screen can be used to quite adequately model most residential development, as is the case here.

The data entry here is relatively self-evident. The development produces 12 apartments with a net area of 40m^2 each. These are expected to sell for £110,000 each. Developer allows this sum to be entered as a unit value, calculating the value per unit of area as a function of this data. The gross area of the apartment block including circulation space is calculated to be used in the determination of the total built area of the residential element (shown in the lower left of the screen).

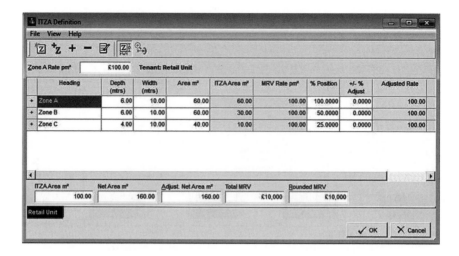

Figure 12.67 ITZA calculation screen

Figure 12.68 The completed commercial area screen for the retail part of the
development

What is less evident is the timing issues. In the earlier developments we have
looked at to date, the timings of the stages has been linear, one has followed
the next without overlap. Developer does, however allow overlapping stages;
indeed it is essential for the accurate modelling of many schemes. Overlapping
is achieved in the timing screen by dragging and dropping the requisite bar,
holding the left mouse button down whilst hovering the cursor over the timing
bar and dragging it to the required place.

With the mixed use development I have done this with the sales period (Figure 12.70).

The reason that this has been done is due to the nature of residential sales. Where residential units are developed they are unlikely to be sold at once but

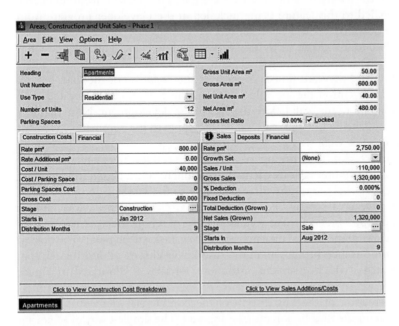

Figure 12.69 Unit sales area for the residential part of the scheme

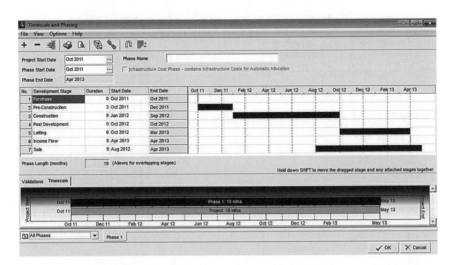

Figure 12.70 The timing for the mixed use development with overlapping timings for the sales period

instead sales will be agreed over an extended period of time. This may well in many projects be before the end of the construction period, certainly this will almost always occur in the development of housing projects and often in the development of blocks such as this one. To model this needs a longer sale period and the ability to distribute sales along it. This demand from residential developers led to this feature being included in the Developer program from version 3 onward.

Care has to be taken with this function, however. The sales are not automatically distributed along the length of the extended sale period, indeed in the template they are timed to occur at the *beginning* of the sales period. Without intervention the site value or profitability would be overstated as all the sales would occur well before the end of the development. Note that in a single phase scheme like this, this would also apply to the commercial element of the project as well.

To correct this, the timing of the sales of both the capitalised rent (commercial) and unit sales (residential) components must be overridden. This is achieved by clicking on the three dots next to the sale stage box in each of the respective area screens. For the residential element, the 'allow custom distribution' box is checked, as well as 'dispose of whole units' (if not then an unrealistic distribution based on a percentage of values will be used). The desired distribution can then be chosen; in this case I have chosen to distribute evenly over the stage, which gives the distribution of sales as illustrated in Figure 12.71.

The correction of the distribution of the commercial element is simpler. Again the template distribution must be adjusted. The simplest way of achieving

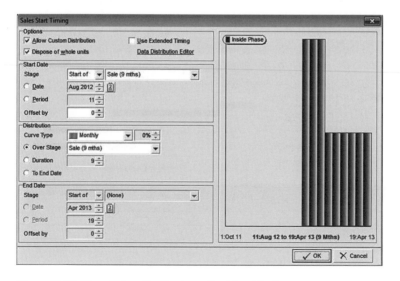

Figure 12.71 Overridden distribution of residential sales

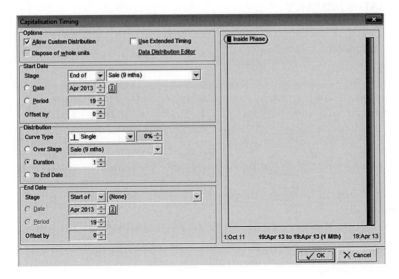

Figure 12.72 Commercial sale timing moved to the end of the sale stage

Heading	9 2012	10 Jul 2012	11 Aug 2012	12 Sep 2012	13 Oct 2012	14 Nov 2012	15 Dec 2012	16 Jan 2013	17 Feb 2013	18 Mar 2013	19 Apr 2013
Revenue											
History - Retail Unit	0	0	0	0	0	0	0	0	0	0	10,000
MRV - Retail Unit	0	0	0	0	0	0	0	0	0	0	10,000
Cap - Retail Unit	0	0	0	0	0	0	0	0	0	0	181,818
Sale - Apartments	0	0	220,000	220,000	220,000	110,000	110,000	110,000	110,000	110,000	110,000
Disposal Costs											
Purchaser's Costs	0	0	0	0	0	0	0	0	0	0	(10,909)
Sales Agent Fee	0	0	(2,200)	(2,200)	(2,200)	(1,100)	(1,100)	(1,100)	(1,100)	(1,100)	(2,809)
Sales Legal Fee	0	0	(2,200)	(2,200)	(2,200)	(1,100)	(1,100)	(1,100)	(1,100)	(1,100)	(2,809)

Figure 12.73 Cash flow extract from mixed-use scheme

this is to again check the 'allow custom distribution' box and then, on the drop down box immediately below, change from 'start of' to 'end of'. This will move the assumed sale from the beginning to the end of the sale (Figure 12.72).

The results of the change in timing can be observed (and checked) using the cash flow (see Figure 12.73). This illustrates the desired distribution with the sales agents fees from the disposal of the units building to the final sale of the freehold of the retail unit.

The appraisal summary for the scheme is illustrated below.

APPRAISAL SUMMARY
Mixed Use Development

Summary Appraisal for Part 1

REVENUE

Sales Valuation	Units	m²	Rate m²	Unit Price	Gross Sales
Apartments	12	480.00	£2,750.00	£110,000	1,320,000

Rental Area Summary				Initial	Net Rent
	Units	m²	Rate m²	MRV/Unit	at Sale
Retail Unit	1	160.00	£100.00	£10,000	10,000

Investment Valuation
Retail Unit

Current Rent 10,000	YP @	5.5000%	18.1818	181,818

GROSS DEVELOPMENT VALUE			1,501,818
Purchaser's Costs	6.00%	(10,909)	

NET DEVELOPMENT VALUE   1,490,909

NET REALISATION   1,490,909

OUTLAY

ACQUISITION COSTS

Residualised Price		436,414	
Stamp Duty	3.00%	13,092	
Agent Fee	1.00%	4,364	
Legal Fee	1.00%	4,364	
Town Planning		5,000	
Survey		3,000	
			466,235

CONSTRUCTION COSTS

Base Construction 800.00 m²	580,000	
Contingency	29,000	
Demolition	25,000	
		634,000

PROFESSIONAL FEES

Architect	4.00%	23,200	
Quantity Surveyor	2.00%	11,600	
Structural Engineer	1.00%	5,800	
Mech./Elec. Engineer	0.75%	4,350	
Project Manager	1.00%	5,800	
C.D. Manager	0.33%	1,914	
			52,664

MARKETING & LETTING

Marketing		10,000	
Letting Agent Fee	10.00%	1,000	
Letting Legal Fee	5.00%	500	
			11,500

DISPOSAL FEES

Sales Agent Fee	1.00%	14,909	
Sales Legal Fee	1.00%	14,909	
		29,818	

Total Additional Costs   3,000

FINANCE

Debit Rate 6.000% Credit Rate 0.000% (Nominal)

Land	25,723	
Construction	10,924	
Letting Void	8,560	
Total Finance Cost		45,207

TOTAL COSTS   1,242,424

PROFIT

                                          248,485

Performance Measures

Profit on Cost%	20.00%	
Profit on GDV%	16.55%	
Profit on NDV%	16.67%	
Development Yield% (on Rent)	0.80%	
Equivalent Yield% (Nominal)	5.50%	
Equivalent Yield% (True)		5.69%
IRR	32.40%	
Rent Cover	24 yrs 10 mths	
Profit Erosion (finance rate 6.000%)	3 yrs 1 mth	

Argus Developer case study five – multiple timings/ single phase

To a certain extent we have already covered these issues within the sections above by implication, however it is worth specifically addressing these issues as they are common.

Essentially what we are dealing with here is where there are multiple elements within a scheme, each of which have their own distinct timing, yet where the project has only one phase. This definition is by its nature rather indistinct; what actually makes a separate phase in a project? Despite this, I believe that this will be understood by most people involved with development projects.

The example we will use is based on the multiple building example used in the section above. As noted, the template used by Developer will model all of the development to take place over the defined construction period. In practice it is unlikely that all of the buildings will be built at the same time.

The schedule of the four types of building are listed in Figure 12.74.

I have expanded the construction window from the original example from 12 months to 15 months (Figure 12.75).

Let us assume that the build schedule and letting schedule shown in Table 12.1 will occur.

RENT & SALES SCHEDULE
Multiple Building Type Example-multiple timing
RENT AND CAPITALISATION

Areas (Sq Metres)	Units	Area/Unit m²	Total Net Area m²	Rent £ m²	Gross MRV £ pa	Adjustment	Net MRV £ pa	Yield %	YP
Mixed Commercial									
Office Type 1	2	880.00	1,760.00	175.00	308,000	0	308,000	6.00	16.6667
Office Type 2	3	660.00	1,980.00	180.00	356,400	0	356,400	6.25	16.0000
Retail Warehouse Type 1	2	1,425.00	2,850.00	230.00	655,500	0	655,500	5.75	17.3913
Retail Warehouse Type 2	2	950.00	1,900.00	240.00	456,000	0	456,000	6.00	16.6667
Totals			**8,490.00**		**1,775,900**	**0**	**1,775,900**		

Figure 12.74 Types of building in development

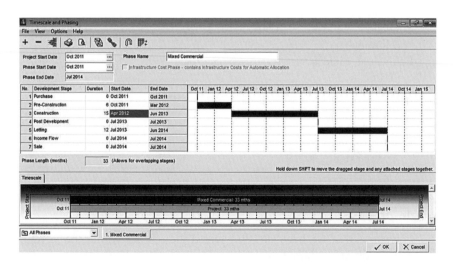

Figure 12.75 The timing screen for the multiple building/multiple timing project

Table 12.1 Build and Letting Schedule

Type	Build Commencement Month	Build Completion Month	Letting Completes
Office Type 1	April 2012	January 2013	January 2014
Office Type 2	November 2012	July 2013	July 2014
Retail Warehouse Type 1	April 2012	October 2012	October 2013
Retail Warehouse Type 2	January 2013	July 2013	July 2014

During the section above on multiple building types I noted that the more that each building type is treated individually, the more flexibility we would have in modelling the project. It is noted that all of these types are each covered in a single area sheet according to their built area. If each building had its separate area sheet then it would be possible to time each individual building separately.

To achieve the desired distribution we need to override the timing assumptions for the construction period and the leasing dates for each of our property types. The original area screen is displayed in Figure 12.76.

To change the construction timing we click on the three dots next to the box titled 'Stage' which has construction highlighted in it. This opens up the timing screen for the construction element (Figure 12.77).

To alter the timing we check the box 'Allow custom distribution'. This opens up the screen for editing. For our purposes we are starting construction in April 2012 and completing at a given date, January 2013. To select the end date we must click on the 'To end date' in the distribution section then select the end

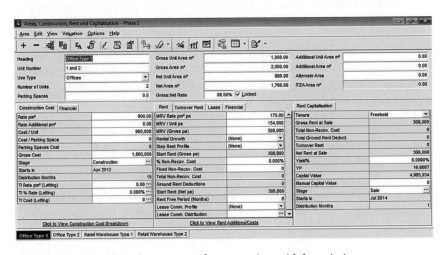

Figure 12.76 The original area screen for our project with base timings

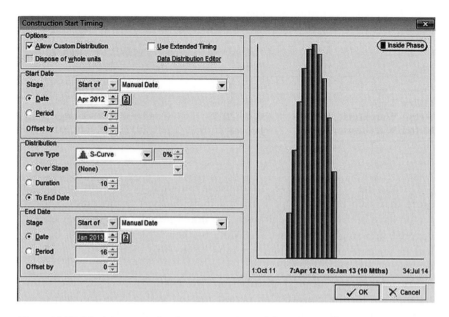

Figure 12.77 Timing screen for the construction of the type 1 offices

date in the bottom section either by using the calendar function or else moving the date box arrows to the required date.

The alteration of the assumed date of letting (which is assumed to occur 12 months after the end of construction hence the need to adjust this timing) is achieved by clicking behind the normal rent screen in the bottom central part of the screen onto the 'Lease' tab (see Figure 12.78).

The timing we have to alter is the lease start date. Again we click onto the three dots to bring up the timing screen. To make alterations to the timing we have to check 'Allow custom timing'. This makes the screen options editable. Selecting 'date' we can then select the date either by the calendar function or by adjusting the up/down arrows.

This process is repeated for the other three property types.

The results of the changes we have made can be observed in the cash flow screen (see Figures 12.79 and 12.80). Alterations can be made from here in addition if required.

Essentially any aspect of the timing of a development project may be adjusted in Developer. What I have tried to illustrate are common areas where developers/appraiser will want to adjust the timing away from that provided in the standard template.

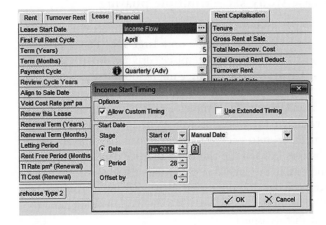

Rent	Turnover Rent	Lease	Financial		Rent Capitalisation

Lease Start Date — Income Flow ⋯ — Tenure
First Full Rent Cycle — April ▼ — Gross Rent at Sale
Term (Years) — 5 — Total Non-Recov. Cost
Term (Months) — 0 — Total Ground Rent Deduct.
Payment Cycle — ⓘ Quarterly (Adv) ▼ — Turnover Rent
Review Cycle Years — — Net Rent at Sale

Income Start Timing ✕

Options
☑ Allow Custom Timing ☐ Use Extended Timing

Start Date
Stage — Start of ▼ Manual Date ▼
◉ Date — Jan 2014 ⬍ 📅
○ Period — 28 ⬍
Offset by — 0 ⬍

Align to Sale Date
Void Cost Rate pm² pa
Renew this Lease
Renewal Term (Years)
Renewal Term (Months)
Letting Period
Rent Free Period (Months
TI Rate pm² (Renewal)
TI Cost (Renewal)

rehouse Type 2

✓ OK ✕ Cancel

Figure 12.78 Alteration of the assumed date of letting on the Lease tab

Heading	4 ar 12	7 Apr 12:Jun 12	10 Jul 12:Sep 12	13 Oct 12:Dec 12	16 Jan 13:Mar 13	19 Apr 13:Jun 13	22 Jul 13:Sep 13	25 Oct 13:Dec 13	28 Jan 14:Mar 14	31 Apr 14:Jun 14	34 Jul 2014
– Construction Costs											
Demolition	0	(75,000)	0	0	0	0	0	0	0	0	0
Road/Site Works	0	(150,000)	0	0	0	0	0	0	0	0	0
Con. - Office Type 1	0	(363,932)	(719,479)	(614,605)	(101,984)	0	0	0	0	0	0
Con. - Office Type 2	0	0	0	(432,169)	(854,381)	(729,844)	(121,106)	0	0	0	0
Con. - Retail Warehouse Type 1	0	(655,590)	(978,496)	(165,914)	0	0	0	0	0	0	0
Con. - Retail Warehouse Type 2	0	0	0	0	(455,271)	(679,511)	(115,218)	0	0	0	0
Contingency	0	(50,976)	(84,899)	(60,634)	(70,582)	(70,468)	(11,816)	0	0	0	0

Figure 12.79 Adjusted construction cash flow (quarterly view)

ARGUS Developer - C:\Program Files\ARGUS Software\ARGUS Developer\Data\mixed building types.wcfx V3.00.004

File Edit Project Definition Cash Flow Cost Codes View Tools Help

Data Check Status ⓘ

Project		Definition		Cash Flow		Summary	

Cash Flow View Cycle Monthly ▼

☐ Show Zero Value Rows | View Order | Show Detailed Cash Flows in Contribution Order ▼

Heading	24 Sep 2013	25 Oct 2013	26 Nov 2013	27 Dec 2013	28 Jan 2014	29 Feb 2014	30 Mar 2014	31 Apr 2014	32 May 2014	33 Jun 2014
– Revenue										
History - Office Type 1	0	0	0	0	308,000	308,000	308,000	308,000	308,000	308,000
History - Office Type 2	0	0	0	0	0	0	0	0	0	0
History - Retail Warehouse Type	0	655,500	655,500	655,500	655,500	655,500	655,500	655,500	655,500	655,500
History - Retail Warehouse Type	0	0	0	0	0	0	0	0	0	0
MRV - Office Type 1	0	0	0	0	0	0	0	0	0	0
MRV - Office Type 2	0	0	0	0	0	0	0	0	0	0
MRV - Retail Warehouse Type 1	0	0	0	0	0	0	0	0	0	0
MRV - Retail Warehouse Type 2	0	0	0	0	0	0	0	0	0	0
R Flow - Retail Warehouse Type	0	0	0	0	0	0	0	163,875	0	0

Profit	4,640,719	Revenue	27,844,287	Interest	-2,758,583	Dev. Yield		7.654%	IRR	
Profit on Cost	20.000%	Cost (excl Int & Fees)	-20,444,984	Interest	-2,758,583	VAT		0	Equity IRR	
Land Cost	-10,617,268	Costs (All)	-23,203,567	Loan Fees		0	Pre-Finance IRR		14.8860%	Return on E

All Phases ▼ | 1. Mixed Commercial
ll phases included in this group | Phase 1 | R2:44 C1:34 | Area-Based Value: MRV History | Hidden | ID: 262

Figure 12.80 Adjusted cash flow for the rental income streams

Argus Developer case study six – multiple phasing

Multiple phases are common in development. Often developers will not want to build out a complete development scheme on a large site. The reasons are legion; over-production can swamp a marketplace, driving down values and increasing vacancy rates. There are huge resource implications too: development is capital intensive as well as requiring the use of physical assets such as plant, labour, materials and management to complete, a developer may not have sufficient resources to complete the whole project at once. There are advantages in splitting a development up into packages; a developer may choose not to build out a stage themselves but sell the site onto another developer and take the short-term value increase in the site value rather than the longer term returns they might achieve from the development itself. Whatever the reason, phasing is very common in all types of development and a development appraisal tool must be able to cope with this aspect.

Argus Developer was designed from the outset to deal with phasing. The program can deal with unlimited phases. These phases can be parallel (i.e. the timing of each phase can be the same), consecutive, over-lapping or with gaps between them. A phased appraisal must be considered carefully however as there are some factors which the appraiser must consider carefully in setting up the program so as to achieve the outcome they desire. These issues will be outlined below.

When a project is multi-phased, Developer creates a tab for each phase and also a merged phase tab which amalgamates the results from all the phases together.

The main issue (and potential pitfall) is in regards to land value/cost. As noted several times, Developer has two basic modes of calculation, profitability and land residual calculation. If the profitability calculation is chosen the appraiser/developer will have to decide whether and how to apportion the land cost between each phase or whether to put the land cost into one of the phases only. In the latter case only the merged phase tab will provide an accurate assessment of the return from the project. If, however, the land cost is apportioned there will be issues with the assumed timing of the purchase for the second and subsequent phases if they are timed to occur later, namely that the land purchase, setting up the timescales in the most obviously intuitive way will assume that the land purchase for the later phases occurs in the future.

A similar problem exists with the land residual calculation. Again you have the option (this time prompted by the programme) to calculate a single land residual or a separate land residual for each phase. Essentially the same timing issues and lack of analytical power issues occur.

This may seem to be a trivial issue but it is in fact very serious. It can lead to either overstating the expected returns from a multi-phase scheme or else overstating the values or the bid price of the land.

These issues can best be illustrated in an example. The development being appraised is an industrial/distribution warehouse scheme in the South East of

England. The industrial part will be developed first followed 19 months later by the distribution warehouse.

The land has cost the developer £9,000,000. This has been apportioned £4,000,000 for the industrial phase and £5,000,000 for the more valuable (and larger) distribution warehouses (Figure 12.81).

The basic set up of the programme is as we have covered previously. The first difference that we will come across is with the timing screen.

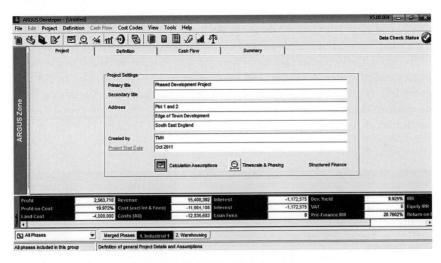

Figure 12.81 Project screen for multiple phase development

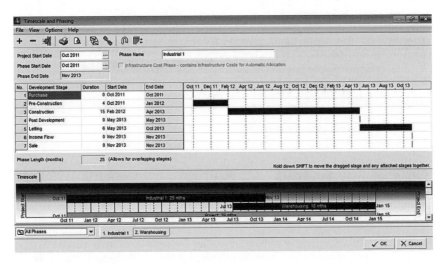

Figure 12.82 Timing screen for first phase

The first phase is created as we have seen previously. The second phase is created by clicking on the green plus symbol. Reference to Figure 12.82 shows that there is a box that allows the definition of a phase start date. The natural inclination, given that we are told that phase 2 starts 19 months after the project commencement, is to set the phase start date as at June 2013 (Figure 12.83).

The creation of a second phase means that three tabs are created on the control panel (Figure 12.84), with both active phases and a merged phase tab, as discussed above. Caution is needed to ensure that data is entered into the correct phase. It is not possible to enter data via the definition tab of the merged phases tab.

The completed definition tabs for phase 1 and phase 2 are illustrated in Figures 12.85 and 12.86. For details of data entry see the earlier parts of this section.

As noted, the definition tab combines the data entered into each of the other phases and provides a consolidated result (Figure 12.87). It will be noted that every box is greyed out meaning that editing is not possible. Alterations must be made in the respective phase definition tab. It is possible, however, to make amendments to the merged phases cash flow. This is because each line in the

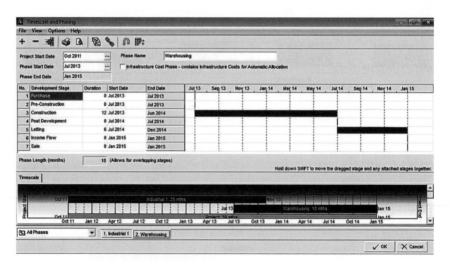

Figure 12.83 Phase 2 timing screen

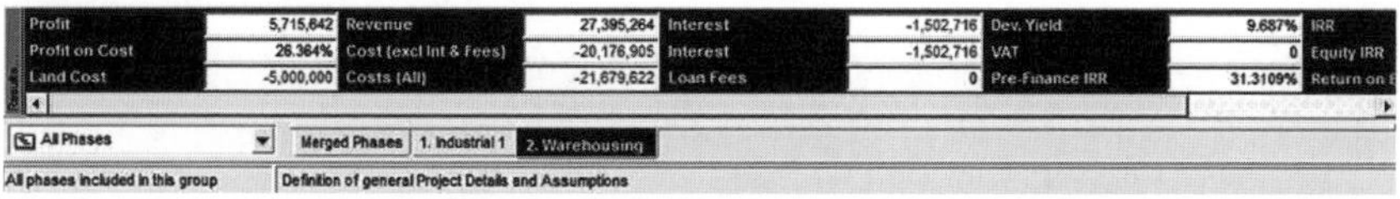

Figure 12.84 Results summary ribbon illustrating the creation of three tabs when a two phase project is created. Note that the warehousing is the active phase

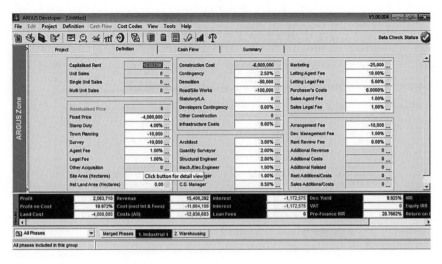

Figure 12.85 Phase 1 definition tab

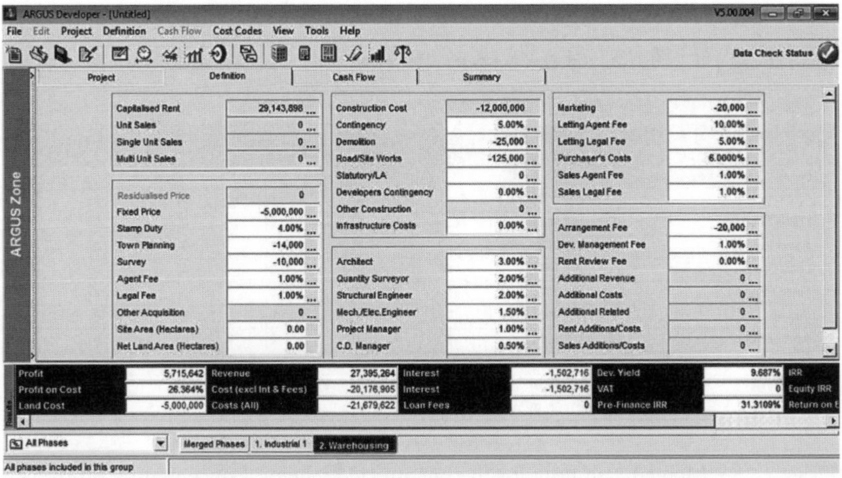

Figure 12.86 Phase 2 definition tab

cash flow is associated with its individual phase so any alteration made in the merged phase cash flow can automatically carry through to the phase cash flow. This is a very useful project management tool.

This produces an apparently viable appraisal, and, if the cash flow was not scrutinised and the appraisal summary used as the appraisal output, the error in the appraisal would be missed. The error becomes apparent only when the merged phase cash flow is viewed (Figure 12.88).

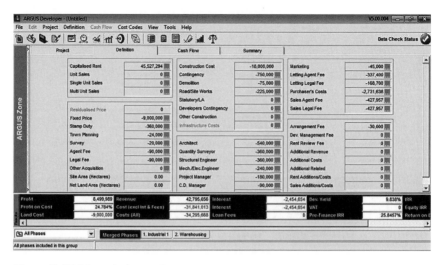

Figure 12.87 Merged phases tab

Heading	1 Oct 11:Mar 12	7 Apr 12:Sep 12	13 Oct 12:Mar 13	19 Apr 13:Sep 13	25 Oct 13:Mar 14	31 Apr 14:Sep 14	37 Oct 14:Jan 15
– Phase 1 - Acquisition Costs							
Fixed Price	(4,000,000)	0	0	0	0	0	0
Stamp Duty	(160,000)	0	0	0	0	0	0
Agent Fee	(40,000)	0	0	0	0	0	0
Legal Fee	(40,000)	0	0	0	0	0	0
Town Planning	(10,000)	0	0	0	0	0	0
Survey	(10,000)	0	0	0	0	0	0
Arrangement Fee	(10,000)	0	0	0	0	0	0
– Phase 2 - Acquisition Costs							
Fixed Price	0	0	0	(5,000,000)	0	0	0
Stamp Duty	0	0	0	(200,000)	0	0	0

Figure 12.88 Merged cash flow for the uncorrected scheme, concentrating on the
timing of land acquisition (semi-annual view cycle)

It can be seen that the programme has split the land purchase both in quantum and time, assuming that the land expenditure for phase 2 occurs in month 19, which is clearly incorrect.

It is obvious that this could have been avoided by not apportioning the land but lumping all of the purchase into phase 1. Then the choice of the phase start date would have had no effect. The downside to this is that it would then be impossible to determine the performance of each phase accurately; only the merged phase sheet would show the true project returns.

The problem can be avoided by using the purchase stage of the timing screen. The phase start date for phase 2 must be returned back to the same as the main project. Then the value of the purchase months can set to the correct 19 month gap (Figure 12.89).

Reference to the merged cash flow shows that this corrects the land purchase bringing both to the same point in time (Figure 12.90).

Note that the profit has changed from just under £8.5m (24.78%) to £7.74m (22.08%), a not inconsiderable change.

The same issues apply when the programme is run in land residual mode. The choice is given to the user to calculate a single land residual or a land

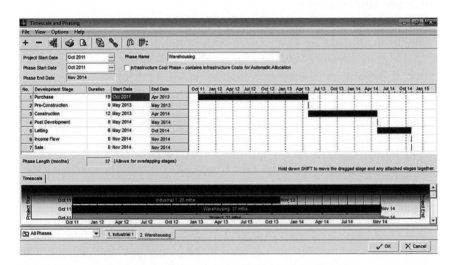

Figure 12.89 Corrected timescale

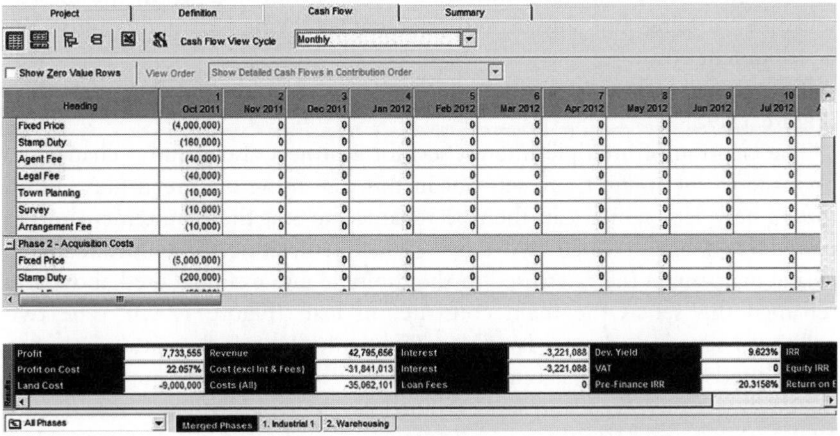

Figure 12.90 Merged phases cash flow of the corrected project

residual for each phase. Unless the land is to be sold/purchased at the beginning of each phase rather than at the project start, the same correction is required to produce an accurate appraisal.

It should be noted that this is not an error in the program, it is simply a common error that can be made with multiple phased projects and an illustration that care must be taken when such projects are being constructed. In fact the multiple phasing mode of Developer is very well thought out and works extremely well.

Argus Developer case study seven – appraisal using multiple interest sets

There is an add on to Developer which allows very sophisticated modelling of the financial make up of a project. This is outside the scope of this book as the advanced finance module changes the characteristics of Developer so much; it would require a separate book to do the topic justice.

You can still do fairly sophisticated financial modelling with the basic version of Developer. One of the elements that can be modelled is the use of multiple interest sets in the program. These sets can be run consecutively or simultaneously.

The former is not really multiple interest sets. It is instead a single, variable rate loan, used either where the loan is scheduled to periodically change by fixed steps or where the loan is on a variable rate and it is expected that the rate will fluctuate. To model this all that needs to change is the rates in the interest set (see Figure 12.91).

Note that I have made two changes. I have introduced a credit rate, the rate of interest that will accrue on any sums if the project goes into surplus. The second is to schedule the differing rates of interest. Note that you have to specify the amount of time (in months) that the rate is expected to continue for. Also the last rate in the schedule should have a 0 against it to avoid the interest schedule running out and a zero percentage being charged against the development costs.

The resulting interest rates will then appear in the finance cash flow (Figure 12.92).

The creation of multiple interest sets for a project also requires changes to be made to the interest rate sets but in this case using the green plus button to add new sets, which can then be renamed to suit the requirement of the user. Here I have created two new sets whilst deleting the empty loan set. I have maintained the rates used in the multiple rate example used above but renamed this set as the main construction loan (Figure 12.93). The two additional sets created are a land loan (Figure 12.94) which is at a single flat, lower rate, something that often can be achieved against land purchase given the lower risks involved, and a mezzanine loan at a typically higher rate (Figure 12.95). (Mezzanine finance is usually arranged to top up existing finance. It is usually short-term, high risk lending on the margins of a scheme.)

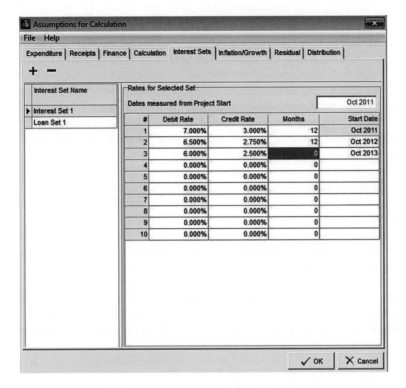

Figure 12.91 Multiple interest rates to model market interest rates movements

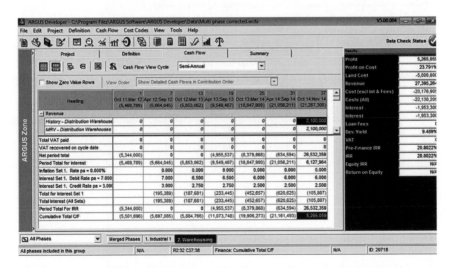

Figure 12.92 Resultant cash flow (semi-annual view and with results bar in alternative position)

Note that with no other changes, Developer would only use the first set of interest rates it finds in this section, in this case the main construction loan. The appraiser has to manually intervene to change the assumptions made.

Most of the schemes use the main construction loan, however the construction in phase 2 (warehousing) is financed via the mezzanine loan, whilst the land is purchased using the separate (and lower) land loan.

As noted above, by default the program already uses the main construction loan by default, so the only changes we need to make is to associate the correct loans with the construction of phase 2 and the land purchase against the land loan. There are two separate techniques involved.

With the construction loan, any items that need to be changed are normally done via the appropriate place on the definition tab or on the area screen. This latter procedure is illustrated with the main construction activity (Figure 12.96). There is a tab behind the construction calculation tab headed 'Finance'. Clicking on this opens up a series of options, including the ability to ascribe an inflation set to this element (also by reference to a library in the assumptions screen). In this case, we want the second box down. This opens up the interest sets and we can select any that we have previously created (Figure 12.97).

The process needed to change items such as the interest set against the land purchase is more involved. Most items in the definition tab have a financial screen when a drill down into the cell detail is made. That for the demolition

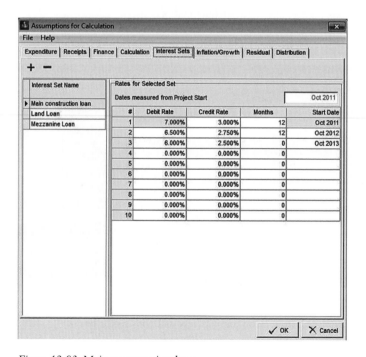

Figure 12.93 Main construction loan

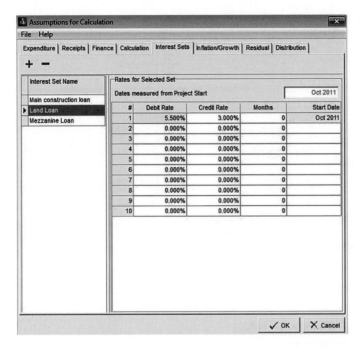

Figure 12.94 Land loan

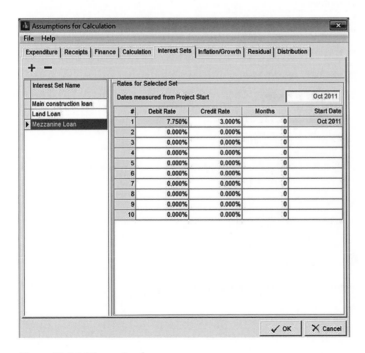

Figure 12.95 Mezzanine loan

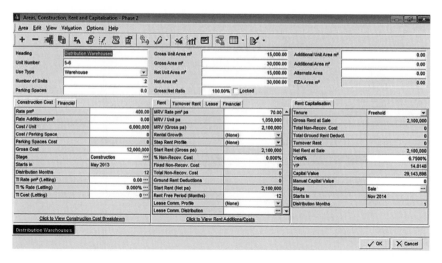

Figure 12.96 Area screen

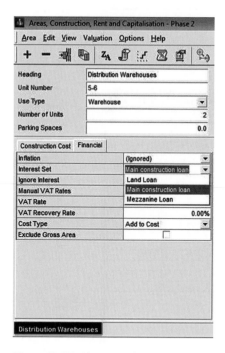

Figure 12.97 Changing the interest set in the financial tab

in phase 2 is illustrated in Figure 12.98. (Note that the icon 'Show financial detail' has to be pressed to make active.)

This option is not available for the two items of land cost. Association with an alternative interest set has to be done via the cash flow (indeed, all changes can be made using the method that is about to be presented). As there are land purchase items in both phases, it is easiest to make the changes in the merged phases cash flow (Figure 12.99).

To access the area that needs to be changed, the user has to click into the appropriate row and then right mouse click to bring up the dialogue box (Figure 12.100).

As can be observed, any item that is not greyed out can be changed. In this case we need to click on Interest and change the interest set to the land loan. Clicking the interest button opens up another dialogue box (Figure 12.101). Once this box is closed the changes will be made and the appraisal total will adjust (Figure 12.102).

Developer does acknowledge when multiple interest sets are used but it is not particularly informative as to the detail (see summary outcome and reports).

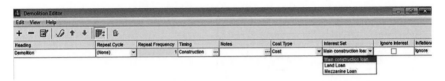

Figure 12.98 Financial detail of demolition item phase 2

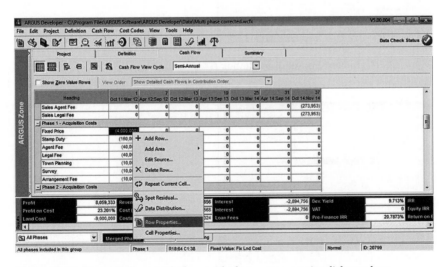

Figure 12.99 Merged phases cash flow with the row properties dialogue box highlighted

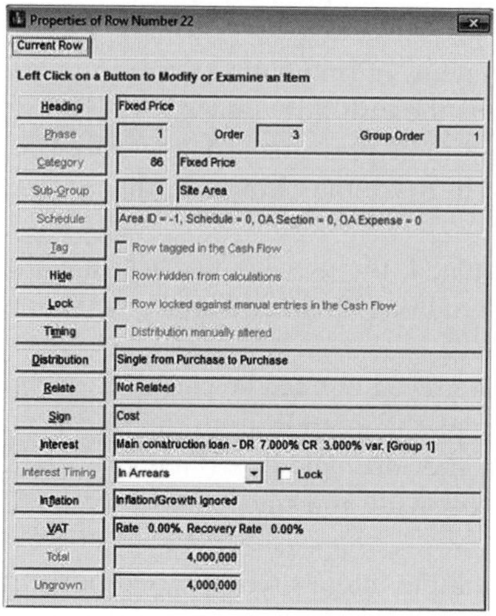

Figure 12.100 Row properties dialogue box

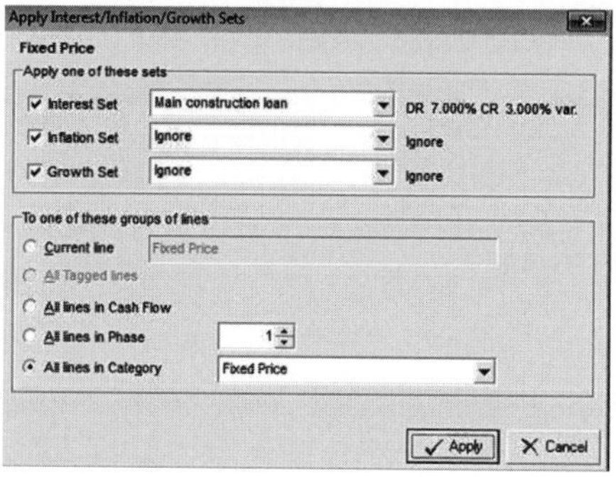

Figure 12.101 Interest button dialogue box

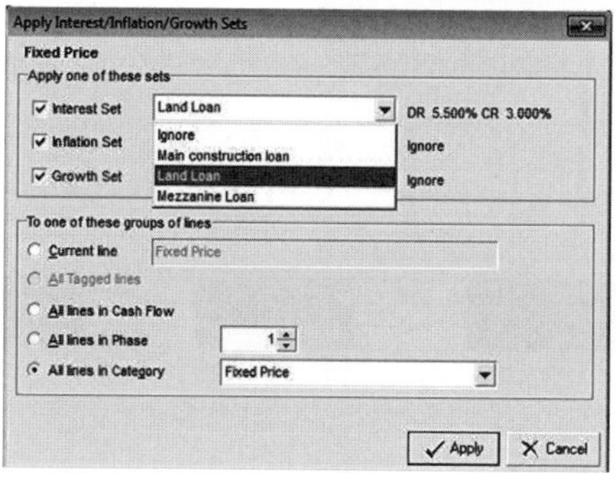

Figure 12.102 Changing the interest set

Argus Developer case study eight – specialist properties: operated assets – golf course development

Operated assets are specialist kinds of commercial property that include hotels, golf courses, marinas and care homes. Although these are diverse property types they have one thing in common; the properties and the businesses that operate them are closely interlinked and the value of these interests are less determined by market rent but by the value of the net income stream a notional, efficiently run business can generate. In the UK market these properties tend to be valued using their own method, the profits method, and development appraisal has been very difficult to conduct on these properties because of their specialist nature. In terms of obtaining market evidence this is still true but Developer has offered a module that makes the physical act of calculating the appraisal of such assets relatively simple.

I am going to run through an example appraisal for the development of a 'pay and play' golf course. Please note that the values used are just for illustrative purposes for the module only and should not be used as an accurate guide. The appraisal has been completed so we will see an analysis of how the values were created (Figure 12.103).

Setting up the appraisal is essentially similar to that done on previous appraisals, therefore we will skip the detail of this, readers who themselves have skipped the earlier sections are referred to the start of section two above. The first change to what is probably the norm in setting up Developer is in regards to the timescale and phasing (Figure 12.104).

This is a clue to the assumptions required is to be found in the name 'operated asset'. This type of property creates its value by being operated and assessing the net income that can be generated after running costs etc. It will probably

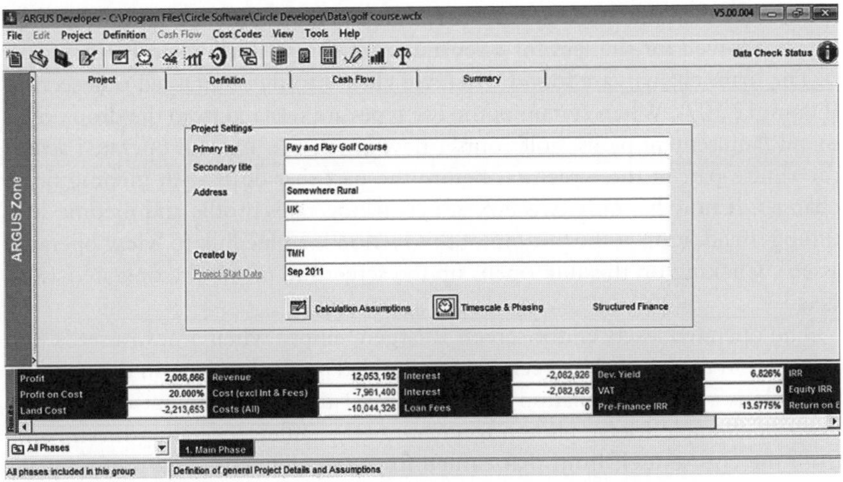

Figure 12.103 Appraisal of 'pay and play' golf course: project screen

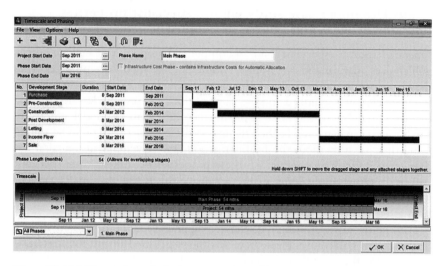

Figure 12.104 Timescale and phasing set up for golf course development appraisal

take time for this business to become established therefore we must model the income and expenditure over a period of time. Essentially what we are working towards is a stabilised net operating income, a phrase that will be very familiar to US appraisers but perhaps less so to UK and European valuers. Whatever, we need to have a period where the operation of the business can be modelled and this is done via development stage 6 of the timescale and phasing sheet, nominally labelled as income flow. I have allowed a 24 month period (Figure 12.105).

The definitions screen is complete, perhaps the only difference from the norm is that the architect's fee is not a percentage but a greyed out box with a sum in it rather than the more normal percentage (Figure 12.106). This is because I have allowed for the fees for a separate golf course designer to be included.

The main changes are found when we click into the capitalised rent section (Figure 12.107). When certain of the use types are selected from the drop down menu (amusement parks, golf course, new hotel, marinas and operated assets) the centre part of the screen, as before the part that deals with income flow, changes. It now has only two boxes; occupancy/rates profile and income start timing. In addition at the bottom of this section is a blue link to 'view operated assets'. Clicking on this link opens up the screens to create the operated assets detail.

The operated assets editor screen is largely empty when a new asset is first created. On the left of the screen is a part entitled profile browser. In this context, a profile is a set of calculation assumptions related to an element of the asset. For this example I have chosen to create two profiles, separating the income from the course itself from that earned from the clubhouse and professional's shop (Figure 12.108).

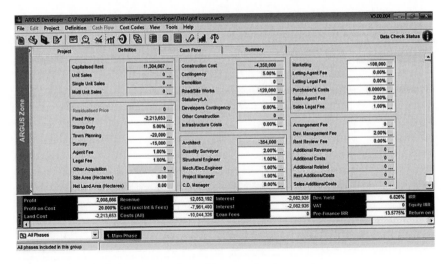

Figure 12.105 Definition screen for golf course development

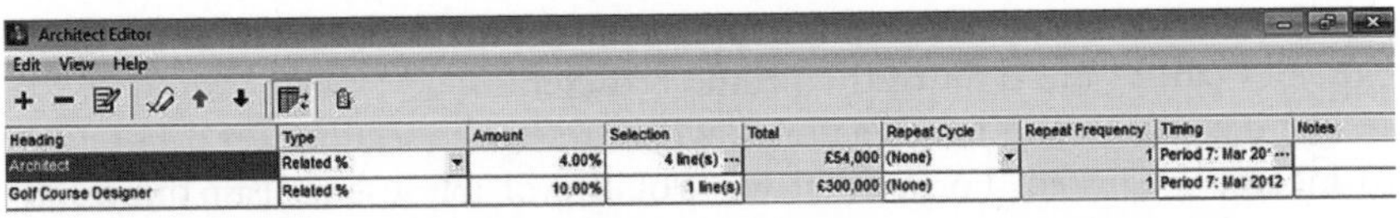

Figure 12.106 Details of architect's fees

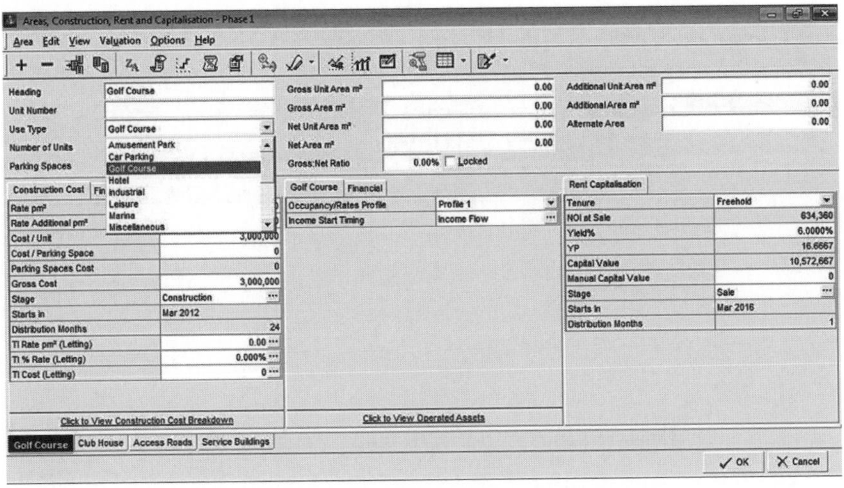

Figure 12.107 The area screen with an operated asset selected

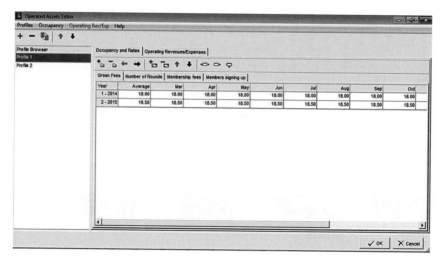

Figure 12.108 The operated assets editor with the two profiles

The main calculation that produces the income forecast from the two profiles is undertaken using the occupancy and rates and operating revenue/expenses tabs to the right of the profile browser.

The occupancy and rates section is where reference sections on data related to income is created. For a hotel we would probably use at least three sub-tabs; for the number of rooms available in each period, the rates for each room (i.e. costs) and the expected occupancy levels in each period. Each of these sub-tabs is created using the first of the plus signs in the tool ribbon above the

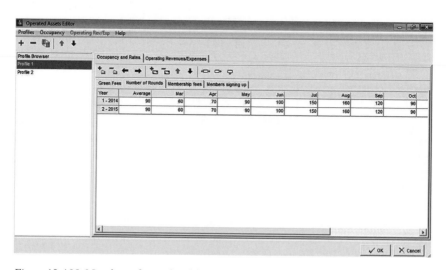

Figure 12.109 Number of rounds sold

section. Additional years of projection are created using the second plus sign to the right. Deletion of unwanted sections are achieved using the appropriate minus buttons.

For the golf course I have created four sub-tabs; for green fees, the number of rounds sold each month, membership fees (although I am modelling a pay and play course, memberships will usually still exist for priority and discounted booking at key times and for handicaps and competitions) and for the project number of memberships sold and renewed each month. These are shown in Figures 12.109 to 12.112.

On the operating revenues and expense tab, the user is required to create and calculate the income and the expenditure for the element. The income side uses the income information we have just defined. In this case we have two elements, the green fees and the membership fees income. Each is calculated

| Occupancy and Rates | Operating Revenues/Expenses |

| Green Fees | Number of Rounds | Membership fees | Members signing up |

Year	Average	Mar	Apr	May	Jun	Jul	Aug	Sep	Oct
1 - 2014	600	600	600	600	600	600	600	600	600
2 - 2015	600	600	600	600	600	600	600	600	600

Figure 12.110 Membership fees

| Occupancy and Rates | Operating Revenues/Expenses |

| Green Fees | Number of Rounds | Membership fees | Members signing up |

Year	Average	Mar	Apr	May	Jun	Jul	Aug	Sep	Oct
1 - 2014	20	20	20	20	20	20	20	20	20
2 - 2015	10	10	10	10	10	10	10	10	10

Figure 12.111 Membership renewals and new sign ups per month

| Occupancy and Rates | Operating Revenues/Expenses |

Revenue and Operating Expense Sections				
Heading	Type	Selection	Detail	Visible
Course Income	Revenue			
Course Expenditure	Expense			
Net income	Section Summary	2 Sections	☑	☑

Operating Revenue and Expense Detail						
Heading	Department Category	Calculation Type	Rate Type	Rate	Selection	Expense Type
Green fees	Course Income	Base Income			2 Pages	··· Revenue
Membership fees	Course Income	Base Income			2 Pages	Revenue

Figure 12.112 Operating revenues and expenses for golf course

by relating the appropriate quantity schedule with the cost schedule, e.g. the cost of each green fee is associated with the expected number of green fees sold using the selection column and checking the appropriate boxes. Both are classified as course income, and both are base income items.

For the course expenditure, i.e. the running costs associated with running the course itself I have created four sections (Figure 12.113). Developer allows a range of calculation techniques but here I have chosen to calculate them as monthly, standalone items. There is an argument that an overall management/ head office cost could additionally be allocated.

Once these two sections have been combined, the net income for this profile is pasted back into the area screen and is used in the capitalisation of the income stream.

A similar calculation process has been followed for the second profile, for the clubhouse (see Figures 12.114 to 12.118). In this case we have two projected income streams, for the bar and food receipts and for the rent for the professional's shop. In this case I have just used simple projections rather than having separate sections for income and quantum. For the expenses I have chosen to calculate these as a percentage of the base income figures, in this case 65 per cent. This is a rather crude rule of thumb and is perhaps not advisable in a real project.

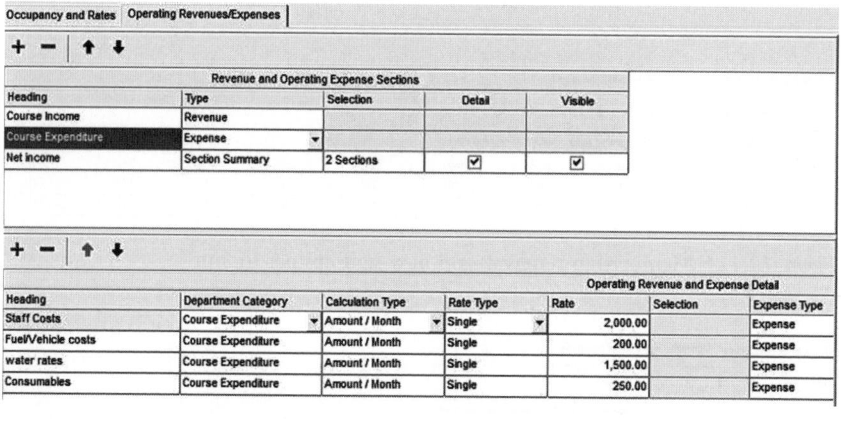

Figure 12.113 Expenditure

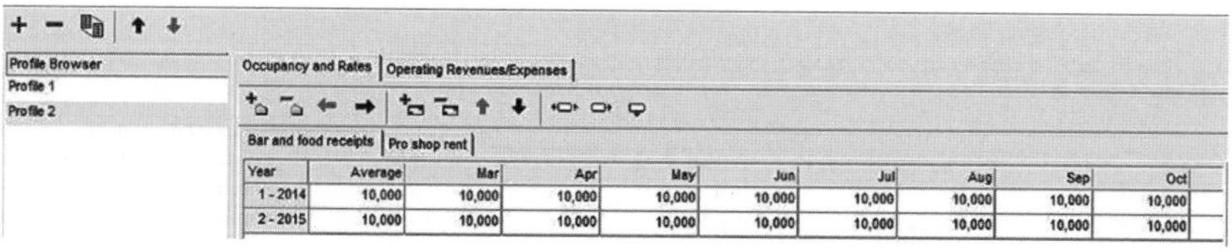

Figure 12.114 Forecast bar and food receipts

Occupancy and Rates | Operating Revenues/Expenses

Bar and food receipts | Pro shop rent

Year	Average	Mar	Apr	May	Jun	Jul	Aug	Sep	Oct
1 - 2014	2,000	2,000	2,000	2,000	2,000	2,000	2,000	2,000	2,000
2 - 2015	2,200	2,200	2,200	2,200	2,200	2,200	2,200	2,200	2,200

Figure 12.115 Professional shop rent

Occupancy and Rates | Operating Revenues/Expenses

Revenue and Operating Expense Sections				
Heading	Type	Selection	Detail	Visible
clubhouse income	Revenue			
operating expenses	Expense			

				Operating Revenue and Expense Detail		
Heading	Department Category	Calculation Type	Rate Type	Rate	Selection	Expense Type
Main Income	Course Income	Base Income			1 Page	Revenue
Pro Shop income	Course Income	Base Income			1 Page	Revenue

Figure 12.116 Revenue stream calculation for clubhouse

Occupancy and Rates | Operating Revenues/Expenses

Revenue and Operating Expense Sections				
Heading	Type	Selection	Detail	Visible
clubhouse income	Revenue			
operating expenses	Expense			

				Operating Revenue and Expense Detail		
Heading	Department Category	Calculation Type	Rate Type	Rate	Selection	Expense Type
Gross operating costs	Course Expenditure	% of Base Income	Single	65.00%		Expense

Figure 12.117 Expenses calculation for clubhouse

Each profile can then be associated with the appropriate development element using the top drop down box in the centre area of the area screen. The income is displayed as the NOI at sale in the right hand section and can be capitalised using an appropriate yield as per a more normal investment. Note that I have also used two more area tabs to calculate the construction cost of non-value elements of the project (Figures 12.119 and 12.120). Developer allows this to be done with ease.

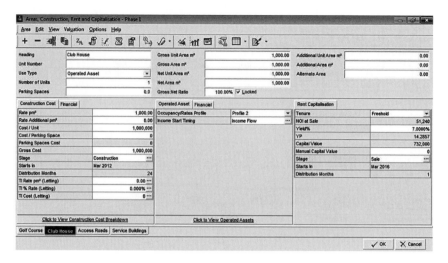

Figure 12.118 Calculation of the value of clubhouse element

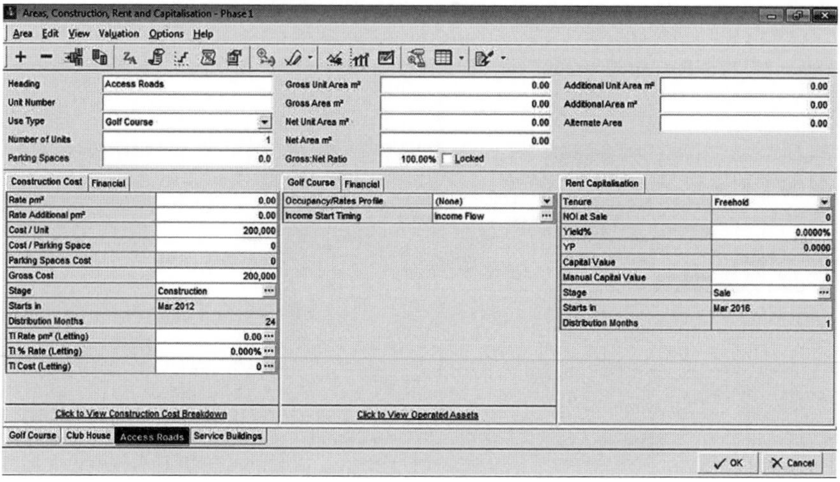

Figure 12.119 Construction of access roads for golf course

The results of the assumptions made in the operated assets sections can be examined and adjusted in the cash flow (Figure 12.121) as has been the case with previous examples. This is a very powerful tool to ensure that the asset has been appraised using the correct assumptions. The cash flow may be imported into Excel; sometimes an Excel view is easier to interpret than the cash flow in the program.

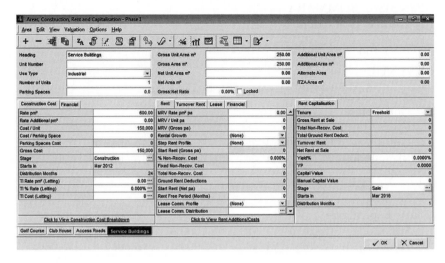

Figure 12.120 Construction of service buildings for golf course

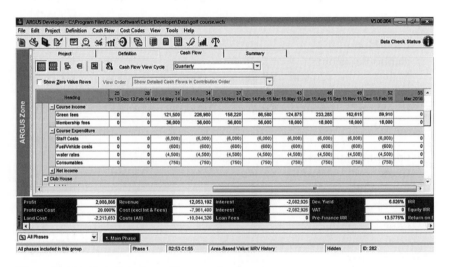

Figure 12.121 Cash flow for golf course development

The appraisal summary (below), lays out the calculation in a traditional fashion.

APPRAISAL SUMMARY
Pay and Play Golf Course

Summary Appraisal for Part 1 Main Phase

REVENUE

Investment Valuation
    Golf Course
    Course Income

Green fees	609,760	YP @	6.0000%	16.6667	10,162,667
Membership fees	72,000	YP @	6.0000%	16.6667	1,200,000
					11,362,667

Course Expenditure

Staff Costs	(24,000)	YP @	6.0000%	16.6667	(400,000)
Fuel/Vehicle costs	(2,400)	YP @	6.0000%	16.6667	(40,000)
water rates	(18,000)	YP @	6.0000%	16.6667	(300,000)
Consumables	(3,000)	YP @	6.0000%	16.6667	(50,000)
					(790,000)
					10,572,667

    Club House
    clubhouse income

Main Income	120,000	YP @	7.0000%	14.2857	1,714,286
Pro Shop income	26,400	YP @	7.0000%	14.2857	377,143
					2,091,429

    operating expenses

Gross operating costs	(95,160)	YP @	7.0000%	14.2857	(1,359,429)
					732,000
					11,304,667

Operated Assets
Golf Course
Course Income

Green fees	1,203,965
Membership fees	216,000
	1,419,965

Course Expenditure

Staff Costs	(48,000)
Fuel/Vehicle costs	(4,800)
water rates	(36,000)

Consumables		(6,000)	
			(94,800)
				1,325,165

Club House
clubhouse income

Main Income		240,000	
Pro Shop income			50,400
			290,400

operating expenses

Gross operating costs			(188,760)
			(188,760)
				101,640

GROSS DEVELOPMENT VALUE			11,304,667
Purchaser's Costs	6.00%	(678,280)	

NET DEVELOPMENT VALUE	12,053,192

NET REALISATION	12,053,192

OUTLAY

ACQUISITION COSTS

Fixed Price		2,213,653	
Stamp Duty	5.00%	110,683	
Agent Fee	1.00%	22,137	
Legal Fee	1.00%	22,137	
Town Planning		20,000	
Survey		15,000	
			2,403,609

CONSTRUCTION COSTS

Base Construction 1,250.00 m^2	4,350,000	
Contingency	217,500	
Road/Site Works	120,000	
		4,687,500

PROFESSIONAL FEES

Architect	4.00%	54,000	
Golf Course Designer	10.00%	300,000	
Quantity Surveyor	2.00%	27,000	
Structural Engineer	1.00%	13,500	
Mech./Elec. Engineer	1.00%	13,500	
Project Manager	1.00%	43,500	
			451,500

MARKETING & LETTING

Marketing	100,000	
		100,000

DISPOSAL FEES

Sales Agent Fee	2.00%	212,528
Sales Legal Fee	1.00%	106,264

318,792

FINANCE

Debit Rate 7.500% Credit Rate 0.000% (Nominal)

Land	466,025
Construction	379,228
Other	1,237,673
Total Finance Cost	

2,082,926

TOTAL COSTS   10,044,326

PROFIT

2,008,866

Performance Measures

Profit on Cost%	20.00%
Profit on GDV%	17.77%
Profit on NDV%	18.90%
Development Yield% (on Rent)	6.83%
Equivalent Yield% (Nominal)	6.06%
Equivalent Yield% (True)	6.30%
IRR	13.58%
Profit Erosion (finance rate 7.500%)	2 yrs 5 mths

In summary, the operated assets module extends the already extensive range that Developer offers the appraiser. It is true that most mainstream users of the program will not need to appraise such specialist projects but the ability to do so is at least there.

Modelling Residential Development Appraisal

The split between residential and commercial development appraisal is a little artificial. Essentially all development appraisal is the same; an appraisal is done to test viability, or to calculate the value of the land for development purposes or a bid price, and this is the same for residential developments as it is for commercial projects.

In reality though, a division is often made. As was noted in the introduction, residential development in the UK was traditionally carried out by different types of developers from commercial schemes; although the division in the industry is now blurred – if not entirely gone – some of the tradition has carried through to the conduct of appraisal. What is more important in the distinction however is the difference in the nature of the product of the residential development process. To generalise, whilst the commercial process tends to produce a small number of buildings which are frequently disposed of in a single sale or over a fairly narrow time frame, residential development produces multiple buildings (or disposable elements in the case of apartment blocks) that have both multiple starts and longer, multiple disposal points.

This latter characteristic produces quite different cash flows from commercial projects and represent enough of a challenge to be considered to be distinct from commercial appraisals. It is these different characteristics which we will be examining in this section of the book. As an aside, it is perhaps instructive to note that many of the principal changes made to Argus Developer over recent years have been to make the program more attuned to producing residential development appraisals reflecting the fact that the program was originally designed to reflect the needs of the market in the late 1980s and early 1990s where the demand for the software came from primarily commercial developers and residential property development was the preserve of the major house builders. As the market moved towards more mixed use and residential development, away from the big house builders, so the program had to change.

Argus Developer case study nine – simple residential project: single building/type/single phase project

The initial appraisal will be undertaken on the simplest of projects; the development of a single detached dwelling on a vacant residential plot in an established residential area. This immediately moves away from one of the characteristics discussed in the introduction – the need to model multiple unit sales over overlapping development and sale periods – however some of the distinctive characteristics of development appraisal will be clear from the appraisal

Land bid calculation

The feasibility study is being done on a residential development plot that was advertised for sale in January 2012 on an internet brokerage site (Figure 12.122).

The main set up assumptions are identical to those used for the commercial and mixed use schemes reviewed earlier in the book. The template Developer uses is flexible enough to be used for both residential and commercial projects. One value difference is in regard to the target profit figure used when a residual land value is calculated (Figure 12.123). Generally the required profit figure for residential developments tends to be lower than for commercial schemes. Essentially this is because the perceived risk for residential projects is lower. It may be worth taking a moment to discuss the reasons for this.

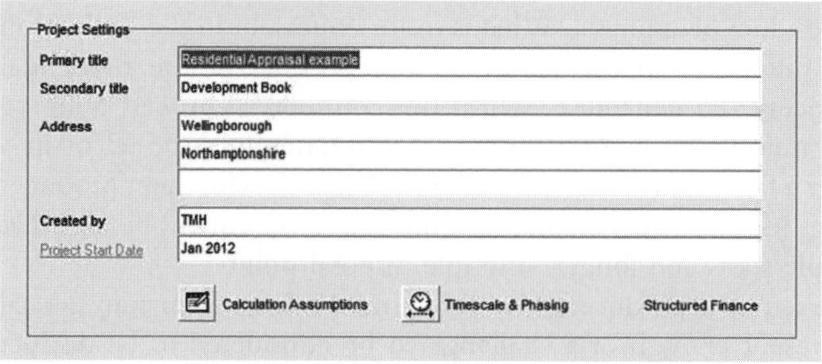

Figure 12.122 Developer opening screen

The lower risk is probably due to the fact that the market for the end product of the development process in residential schemes is wider than for any commercial project. With any commercial project – offices, retail, industrial or leisure – the number of potential end users (occupiers and owners) will naturally be more restricted because there are simply less businesses than households in any market. One of the features of successful speculative development (i.e. development projects that do not have a specific end user in mind at inception) of any type is to ensure that the end product is not so specialised that it excludes all but a very few buyers/occupiers from considering it. This is particularly true of commercial projects. One of the criticisms often

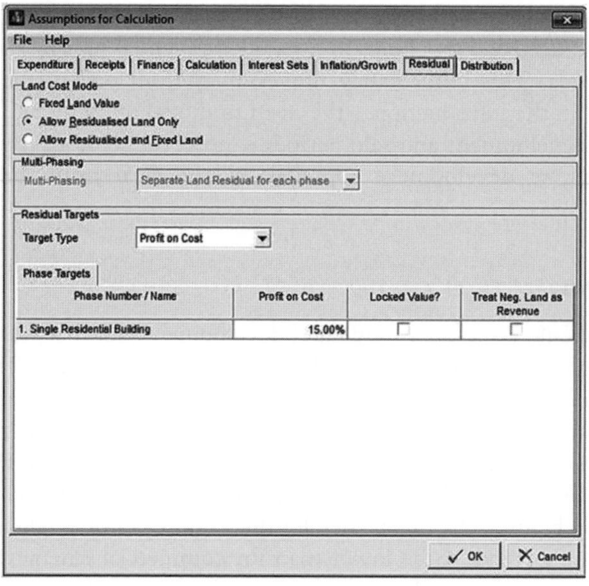

Figure 12.123 The residual assumptions

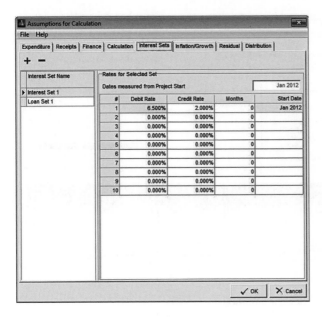

Figure 12.124 Interest sets assumptions

levelled at developers is that their output is bland and unadventurous – sadly bold and innovative designs limit the market appeal and raise development risks to a level that is unacceptable to both developers and their funders alike.

The interest rate (Figure 12.124) used in this appraisal is indicative of the rates that a residential developer with a good track would probably obtain at the time of writing (early 2013). Note that a credit rate has been allowed; this will allow interest to be accrued if the project goes into surplus. This is not something that is likely to happen with this single building project but is something that can occur in multiple unit projects where sales occur over an extended time period and where the crossover between deficit and surplus almost certainly will occur before the final unit is sold and the development project can be assumed to have ended.

As noted this is a simple project that is assumed to start 3 months after land purchase and have a 4 month construction period. I have, however, assumed that there could be a full 6 months after the end of completion for the developer to find a buyer and agree a sale (Figure 12.125). This is a reflection of the uncertainty of the market at the time of writing, however some allowance for this should be made in most appraisals as it is an extra allowance for risk.

There is an issue with the assumed timing of the sale in Argus Developer's default template which will be discussed below.

As previously noted, Developer has three Area Sheets that cover different types of residential development. For this we will use the single unit sales sheet

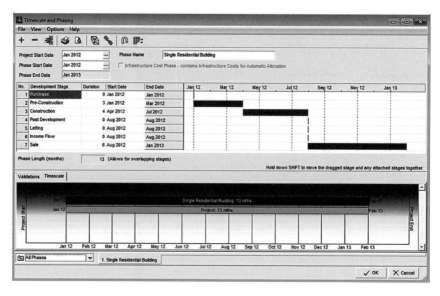

Figure 12.125 Project timescale assumptions

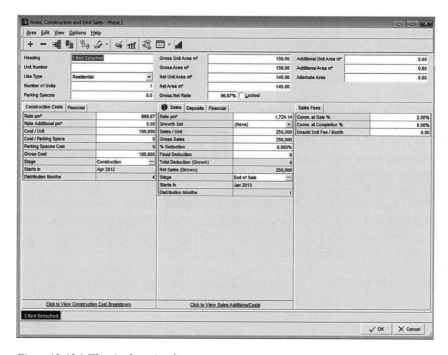

Figure 12.126 The single unit sales area

(Figure 12.126), the one that has been included in the program since the inception of the Windows-based version of the program. This can be seen to be a simplified version of the commercial area sheets, lacking the capitalisation of income area – something which is obvious as a sale of the house or flat is assumed to be the disposal method. (Note that even with investment residential property, it is rare that the investment method is used to determine the value.)

Note that it is normal to enter sale values (and often construction cost) as a 'per unit' rate rather than an area rate.

As noted above, there is a need to amend the timing from the default that exists in the template when an extended time period is used in Argus Developer. The template defaults the timing of the sale to the beginning rather than the end of the period. I believe that this is the case because the programmers believed that if an extended time period were selected then the sales would be multiple and need to be distributed across the full time range. It can also be avoided if one of the post construction time categories is chosen; however I believe that most users would prefer to use the option that I have chosen and that this timing issue is a potentially dangerous trap for the unwary (it will inflate the value/development profitability if it is not spotted). It can be corrected by clicking into the sales timing link on the Area Sales sheet, checking 'allow custom distribution' and timing the start date to the end rather than the beginning of the period (Figure 12.127).

The definition tab for a residential scheme (Figure 12.128) is largely similar to those we have seen previously. Quite often there are fewer members of the professional team, indeed some projects may only have an architect and project manager, or, often, just a project manager where the design has been bought in. Similarly there is no need for letting agent and letting legal fees unless an investment vehicle is being created. Here, the only disposal fees are those for the sale agent.

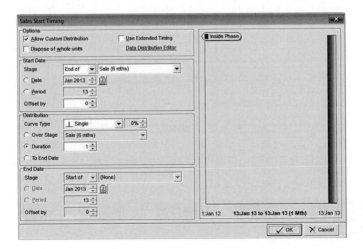

Figure 12.127 Sales construction timing

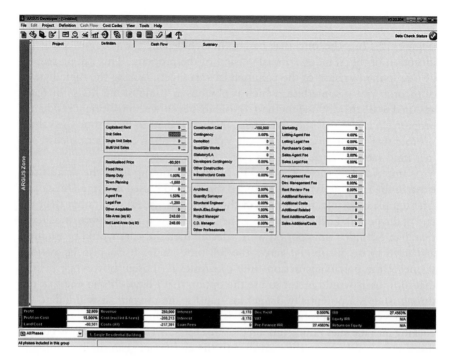

Figure 12.128 Definition tab

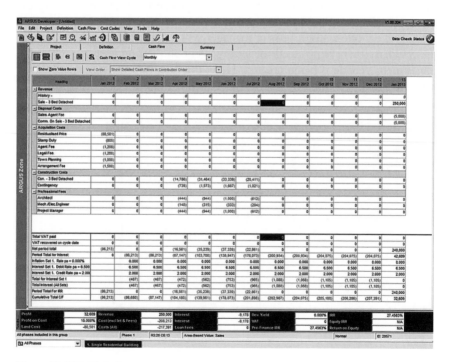

Figure 12.129 Cash flow

The cash flow is relatively simple but provides a useful visual tool for assessing whether the key assumptions have been timed to occur as anticipated (Figure 12.129).

The projects appraisal summary is presented below.

APPRAISAL SUMMARY
Residential Appraisal example

Development Book

Summary Appraisal for Part 1 Single Residential Building

REVENUE

Sales Valuation	Units	m^2	Rate m^2	Unit Price	Gross Sales
3 Bed Detached	1	145.00	£1,724.14	£250,000	250,000

NET REALISATION   250,000

OUTLAY

ACQUISITION COSTS

Residualised Price		80,501	
(248.00 m^2 £324.60 pm^2)			
Stamp Duty	1.00%	805	
Agent Fee	1.50%	1,208	
Legal Fee		1,200	
Town Planning			1,000
			84,713

CONSTRUCTION COSTS

Base Construction			100,000
150.00 m^2 @ £666.67 pm^2			
Contingency		5,000	
			105,000

PROFESSIONAL FEES

Architect	3.00%	3,000	
Mech./Elec.Engineer	1.00%	1,000	
Project Manager	3.00%	3,000	
			7,000

DISPOSAL FEES

Sales Agent Fee	2.00%	5,000	
3 Bed Detached		5,000	
		10,000	
Total Additional Costs			1,500

FINANCE

    Debit Rate 6.500% Credit Rate 2.000% (Nominal)

Land	2,748	
Construction	937	
Other	5,493	
Total Finance Cost		9,178

TOTAL COSTS   217,391

PROFIT

                                            32,609

Performance Measures

Profit on Cost%	15.00%
Profit on GDV%	13.04%
Profit on NDV%	13.04%
IRR	27.46%
Profit Erosion	2 yrs 2 mths
(finance rate 6.500%)	

Argus Developer case study ten – more complex residential projects

The second case study is of a more complex project which more illustrates the characteristics of residential development and its associated modelling issues. It is, indeed, these characteristics which the traditional residual models struggle to cope with so it is instructive to examine these in more detail.

Many residential developments are multi-phased – we have already come across this characteristic when we looked at commercial schemes – however the fact that residential units can be developed and sold in smaller, discrete parcels (i.e. individual houses and flat units) means that each phase can comprise of a series of starts and completions *within* the phases.

The problems that this poses for the traditional model should become apparent. Residual appraisals struggle to adequately model multi-phased schemes and they are simply too broad a tool to deal with this additional complexity. Adopting a cash flow approach is really the only solution, and, of course, although time consuming, an Excel model can be developed relatively easily to deal with these issues. Surprisingly, however the older versions of Developer also had issues with dealing with this specific problem, which perhaps reflects its roots in commercial development appraisal. This was corrected from Version 4 of the program onwards (released in mid-2008) and it is this which forms the basis of the modelled development below. It should be noted that you can model developments with these characteristics in the older versions of the program but there is more work involved, and this will be referred to in the text below.

The project which is modelled over the next pages is a two phase residential development project in Cheshire. There are two elements to the project; firstly the conversion of existing agricultural buildings (barns and outbuildings) into duplex flats and, secondly, a new build phase of detached houses. These two individual elements form the individual phases but both have the 'mini-phases' staged starts and completions within the phases (Figure 12.130).

The basic set up and assumptions is as we have covered before. The basic interest sets have been used applying a loan rate that is relevant to market conditions at the time of writing (early 2013). The timescale and phasing section is worth commenting about; firstly, although it is intended that Phase 2 will start 10 months after project commencement, this has been achieved by using an extended preconstruction stage for phase 2 (see Figures 12.131 and 12.132) with both stages start dates being simultaneous. This is to cover the issue discussed in the multi-phased commercial section above, namely to achieve an accurate project return/land valuation assessment by ensuring that the land purchase for each phase occurs in the correct timeframe. The second thing to say about the timescale and phasing definition is that, unlike what we have seen previously, for key elements in the project (the construction and sale of the units) the timing is *not* actually determined here. Note, however, that all other timings of the other element of the project are set here so the timescale still needs to be defined.

The big difference in the appraisal is found in the definition tab. Instead of using the first or second of the four buttons in the upper left hand part of the definitions screen we use the third, the single unit sales option. Prior to version 4 of the programme only the first two buttons were available so readers with older versions of Developer will find that they do not have these options.

Pressing the button for the first opens up a dialogue box (Figure 12.133) that invites the user to create a new area screen, and this itself is an indication that a new area of the program is being opened up, and indeed it is important to stress once again that it is essential to redefine elements of the timescale in this screen. This MUST be done because selecting this option means that the construction costs and sales data will not be distributed as per the underlying

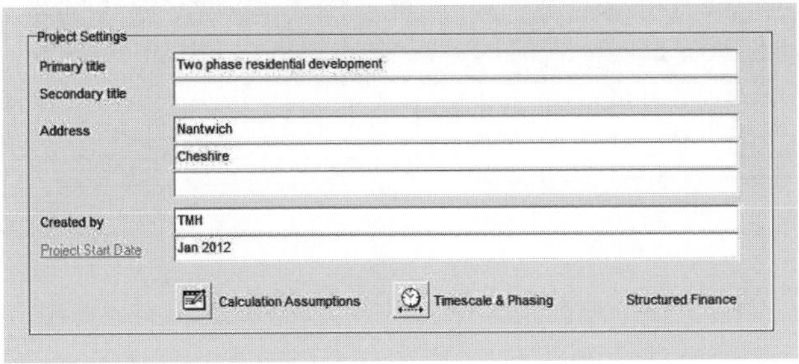

Figure 12.130 Project screen for more complex project

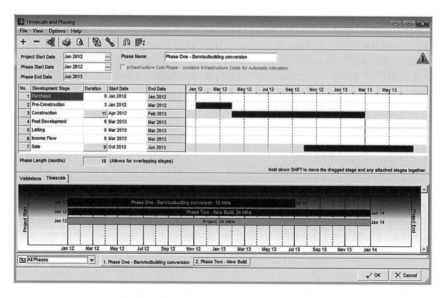

Figure 12.131 Stage one timescale

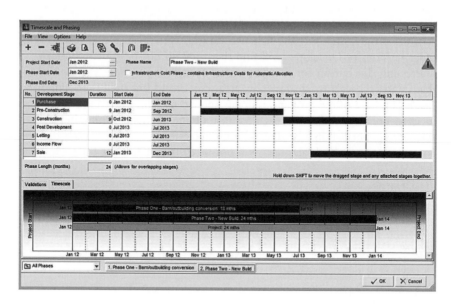

Figure 12.132 Stage two timescale

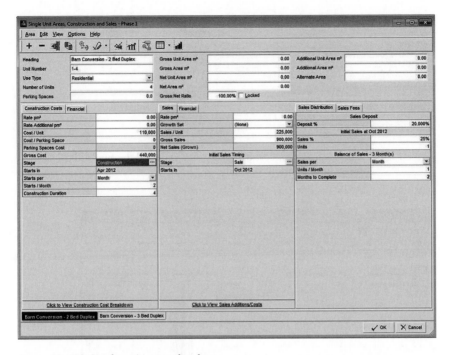

Figure 12.133 Single unit area sales sheet

template but instead will default to single month expenditure and receipts which will, of course, greatly distort the outcome of the appraisal. I am stressing this point because regular users of Developer will be used to its largely fool-proof and transparent characteristics and may be lulled into a false sense of security because of this.

What makes this worse is that the single unit sales area is quite similar in appearance to the Unit area sales sheets and a user in a hurry could confuse the two. The mistake would become obvious if the cash flow was examined but this could be rather too easily overlooked.

Where the differences come in between the two sheets are found in the data entry boxes for construction costs and sales.

For the former, at the bottom of the construction cost area are two boxes – starts/month and construction duration (Figure 12.134). These boxes must be filled in to model the construction expenditure correctly. This function creates a nested series of s-curved expenditure profiles. Here I have assumed that a duplex is started every 2 months and that the construction work on each takes 4 months.

The distribution of the construction costs generated by what are effectively four mini projects can be seen in Figure 12.135.

Construction Costs	Financial	
Rate pm²		0.00
Rate Additional pm²		0.00
Cost / Unit		110,000
Cost / Parking Space		0
Parking Spaces Cost		0
Gross Cost		440,000
Stage	Construction	•••
Starts in	Apr 2012	
Starts per	Month	▼
Starts / Month		2
Construction Duration		4

Figure 12.134 The construction cost data entry area showing the two additional boxes at the bottom of the list

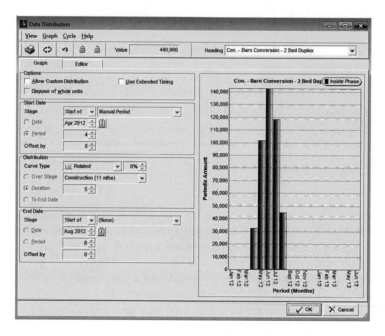

Figure 12.135 Construction cost distribution generated

A similar procedure is followed for the sales of the completed units and, once again, it is stressed that if the Single Unit Sales option is chosen these entries must be made. There are five input areas that are unique to this section and entry into four of them is mandatory. The first mandatory entry is to define when the sales are to start, although this will default to the start of the sales period defined in the timescale and phasing assumptions. The second are that needs to be defined is the number of units that are expected to have been sold at the commencement of the sales period. This is to allow for units sold 'off-plan' or agreed during the construction or pre-construction period. The final two mandatory areas are the number of units expected to be sold each month and the length of time each sale will take to complete. The optional entry is in regards to the deposit. If an assumption is made then the program will model the deposit being made when the sale is agreed and the balance being paid on the expected completion.

The assumptions made for phase one of this project are illustrated in Figure 12.136.

The distribution of these individual sales can be examined using the graphical function that was also developed for this module (see Figure 12.137 which also includes the sale assumptions for phase 2).

Sales	Financial			Sales Distribution	Sales Fees	
Rate pm²			0.00		Sales Deposit	
Growth Set	(None)		▼	Deposit %		20.000%
Sales / Unit			225,000		Initial Sales at Oct 2012	
Gross Sales			900,000	Sales %		25%
Net Sales (Grown)			900,000	Units		1
	Initial Sales Timing				Balance of Sales - 3 Month(s)	
Stage	Sale		···	Sales per	Month	▼
Starts In	Oct 2012			Units / Month		1
				Months to Complete		2

Figure 12.136 Sales structure assumptions

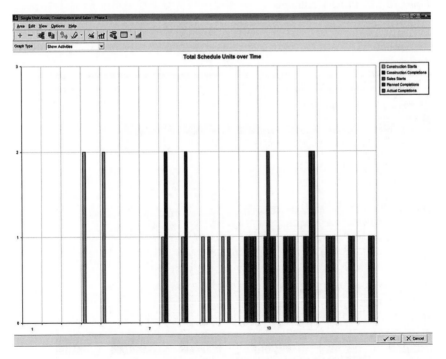

Figure 12.137 Sales distribution for Phases 1 and 2 which results from the assumptions made within the single unit sales area screen

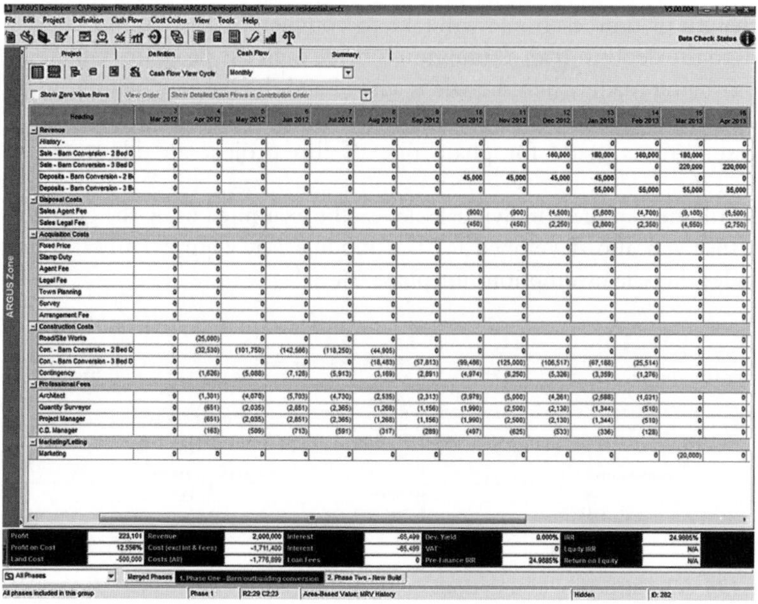

Figure 12.138 Phase 1 cash flow

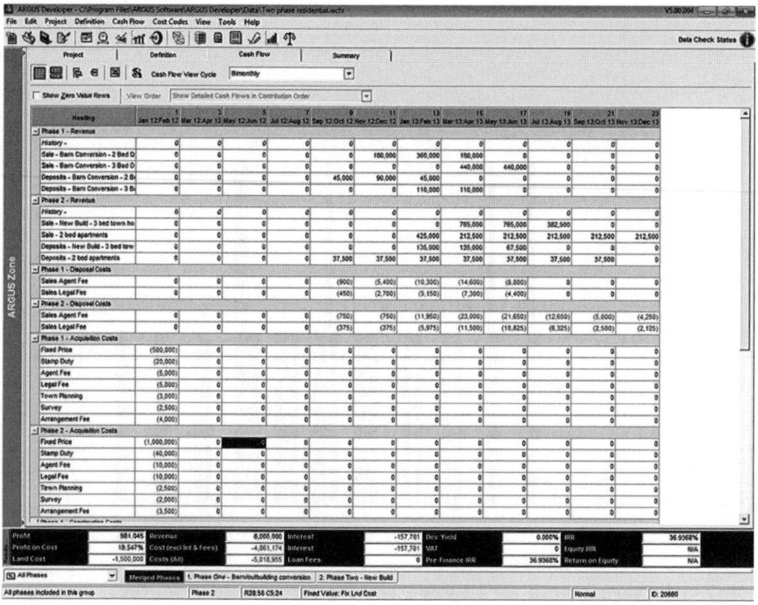

Figure 12.139 Phase 2 cash flow

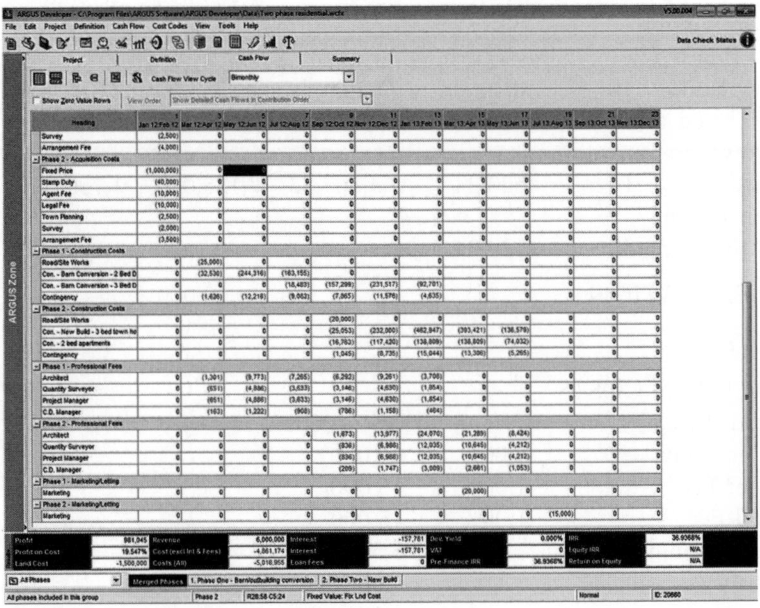

Figure 12.140 Merged phases cash flow

As noted, it is possible to achieve virtually the same results with the unit area sales and with older versions of the programme. To do this each individual unit (duplex or house) would have to have their own area screen and each one would then have to have its timing individually defined. Whilst this is possible with small developments this would become onerous for larger projects.

The results of the assumptions made are best seen in the cash flow screens (Figures 12.138 to 12.140). The data distribution can be amended directly from these screens if required.

The development can then be summarised via the Appraisal Summary screen and print output which lays out the project in the traditional residual format.

APPRAISAL SUMMARY
Two phase residential development

Summary Appraisal for Merged Parts 1 2

REVENUE

Sales Valuation	Units	m^2	Rate m^2	Unit Price	Gross Sales
Barn Conversion – 2 Bed D	4	0.00	£0.00	£225,000	900,000

Barn Conversion – 3 Bed D	4	0.00	£0.00	£275,000	1,100,000
New Build – 3 bed town ho	10	0.00	£0.00	£225,000	2,250,000
2 bed apartments	14	520.24	£3,363.83	£125,000	1,750,000
Totals	32	520.24			6,000,000

NET REALISATION   6,000,000

OUTLAY

ACQUISITION COSTS
Fixed Price		1,500,000
Stamp Duty	4.00%	60,000
Agent Fee	1.00%	15,000
Legal Fee	1.00%	15,000
Town Planning		5,500
Survey		4,500
		1,600,000

CONSTRUCTION COSTS
Base Construction 585.34 m² @ £830.00 pm²		2,675,832
Contingency		90,396
Road/Site Works		45,000
		2,811,228

PROFESSIONAL FEES
Architect	4.00%	107,033
Quantity Surveyor	2.00%	53,517
Project Manager	2.00%	53,517
C.D. Manager	0.50%	13,379
		227,446

MARKETING & LETTING
Marketing	35,000
	35,000

DISPOSAL FEES
Sales Agent Fee	2.00%	120,000
Sales Legal Fee	1.00%	60,000
		180,000
Total Additional Costs		7,500

FINANCE
Debit Rate 6.750% Credit Rate 2.000% (Nominal)
Total Finance Cost   157,781

TOTAL COSTS	5,018,955
PROFIT	
	981,045

Performance Measures

Profit on Cost%	19.55%
Profit on GDV%	16.35%
Profit on NDV%	16.35%
IRR	36.94%
Profit Erosion (finance rate 6.750%)	2 yrs 8 mths

As noted, whilst it is possible to model these types of projects in the older versions of Developer (and in Excel), the single unit sales module does make the process of modelling complex residential projects with quite detailed and sophisticated assumptions relatively easy.

Note that the fourth option within the program, multiple unit sales, allows the modelling of the construction and sales of blocks of residential units, something that is rarely encountered in the UK market. Essentially though, the procedures followed are similar to those outlined in this section.

Argus Developer case study eleven – modelling risk in Argus Developer

The sensitivity analysis module in Argus Developer

Readers are referred to the section on risk and sensitivity in Chapter 9.

To explore how basic sensitivity analysis is done in Argus Developer we will return to some of the simpler projects we examined earlier (Figure 12.141).

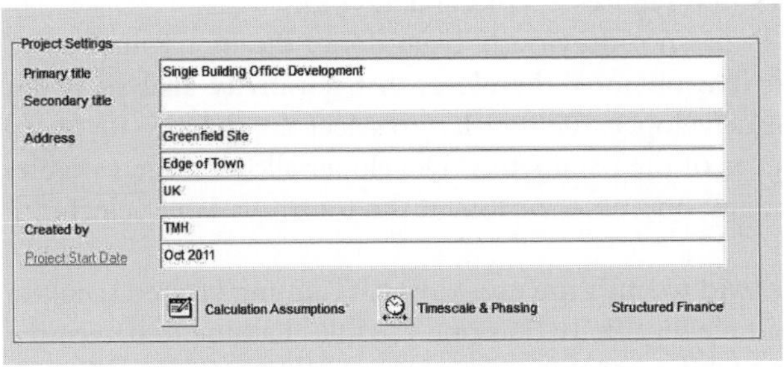

Figure 12.141 Simple commercial project

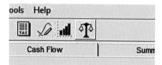

Figure 12.142 Accessing the sensitivity analysis module in Argus Developer

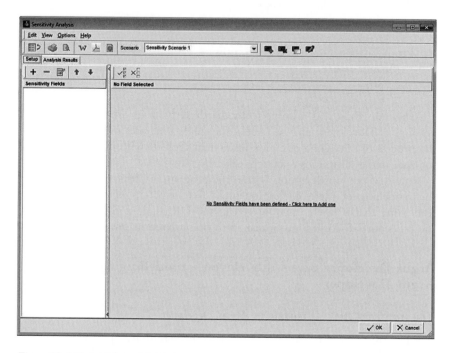

Figure 12.143 Initial sensitivity screen

Dealing with Argus Developer first, the sensitivity analysis module is accessed either from the drop down menu (tools) or via the dedicated button on the toolbar, the set of scales (see Figures 12.142 and 12.143).

If a sensitivity analysis has not previously been done, the user is invited to create sensitivity fields. It is obvious, therefore, that sensitivity analysis is not an automatic feature in Developer; the user has to select the fields to be tested and then define the values of the parameters. Developer allows three variables to be tested together at any one time, or four if the fourth variable selected is time.

These characteristics tend to push the user towards running limited scenario type analysis. Strictly, simple sensitivity tests one variable at a time to determine the impact on returns or values, and this can be done with Developer, although it is invariably the same few factors which always impact on the outcome of the appraisal; anything to do with value – rents, yields, sale values etc. – and it is these which the appraiser needs to pay the greatest attention to.

The chance of single variables changing in isolation is limited; variables inevitably will tend to move together. For example, in a downturn, the project duration tends to increase because the developer finds it harder to find tenants or agree sales of units. At the same time rental values tend to fall because of the fall in demand. As rents fall so do capital values, not only because of the reduced income but also because yields increase as investors seek higher initial returns to compensate them for increased risk and lower future rental growth. On the upside, construction costs tend to fall in these conditions as contractors compete more strongly for what little work is available (although of course this can only apply to construction work for which contracts have not yet been awarded).

The nature of the development environment thus suits scenarios however Developer is slightly limited by allowing only the three variables to be compared at any one time. The variables I have chosen to test are construction costs, rental value and yield – the three most important 'physical' factors in most developments.

The first step in running the analysis is selecting the sensitivity field. This then gives the user the choice of components within that field to be tested (Figure 12.144).

Once this has been chosen a further screen appears (Figure 12.145). This is where the element to be tested is selected for inclusion within the sensitivity analysis (here we have only one element – i.e. a single building but when multiple buildings are included it gives the user the option to test the viability against a single element. It would be normal practice to select every element for analysis). The step type is chosen (fixed or percentage change in values), the step intervals, and whether the analysis is to be unidirectional (i.e. just the

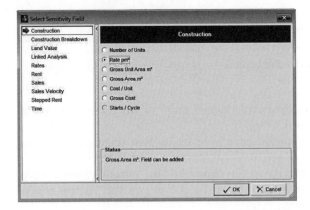

Figure 12.144 Selection of construction element and the component (cost per m²) to be tested in the sensitivity analysis

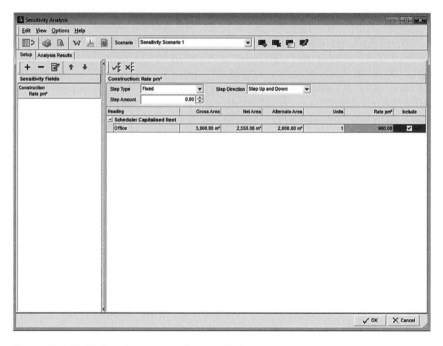

Figure 12.145 Setting the construction sensitivity parameters

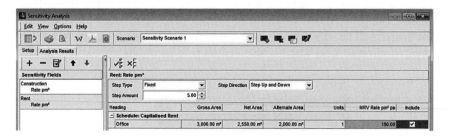

Figure 12.146 Rent sensitivity and options

sensitivity to upward or downward changes only) or whether it is to be up and down analysis (Figure 12.146).

Similar sensitivity assumptions have been set for rent (Figure 12.147) and yield (Figure 12.148).

Once the sensitivity fields have been chosen and the variable values defined, the sensitivity analysis can be run, either by clicking on the 'analysis results' tab or using the drop down menus.

The results of the sensitivity analysis are expressed in tabular form (Figure 12.149). When a third (or fourth (time)) variable is included the

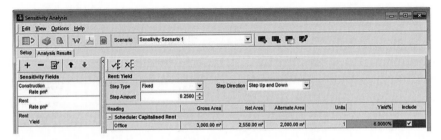

Figure 12.147 Rent sensitivity set up screen

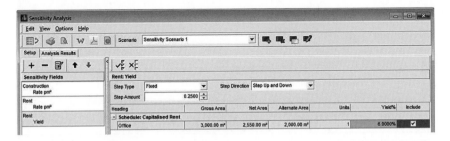

Figure 12.148 Yield sensitivity set up screen

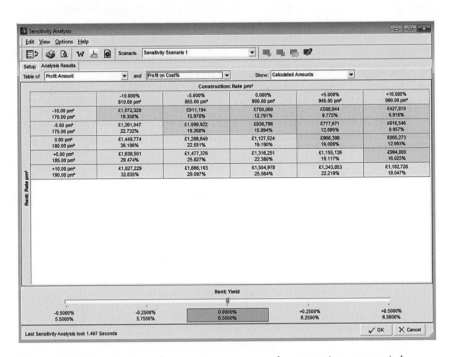

Figure 12.149 Sensitivity analysis outcome – rent and construction costs varied – yield static

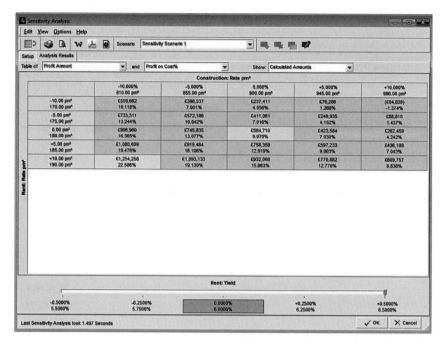

Figure 12.150 Sensitivity analysis outcome – varying the yield using the slider
function

sensitivity impact of this is explored using the slider function at the bottom of
the table (Figure 12.150). All tables can be printed and reported.

In conclusion, the sensitivity function in Argus Developer is relatively
comprehensive and easy to use and links well with reporting. The scope of
the analysis is, however, limited by the number of variables that can be tested
together.

13 Modelling development financial feasibility in Estate Master DF – software outline and case studies

An overview of Estate Master DF

The last example of case studies using proprietary development appraisal software I will examine is Estate Master Development Feasibility, or DF.

The company was founded in 1991 in Sydney, Australia and although Australia and New Zealand are still perhaps its main markets the software is distributed worldwide with users in Australia, New Zealand, the UK, Eire and the UAE. It has over 1000 companies licenced to use its products with an estimated 10,000 individual users. Both in terms of the number of users and its global spread, this is lower than its big rival Argus Developer but is clearly still very significant.

Estate Master DF sits alongside a suite of other software; Development Management DM, Estate Master IA (investment appraisal), Hotel Feasibility HF and Corporate Consolidation CC, the latter allowing portfolios of projects to be amalgamated for analysis. For the purposes of this book we will concentrate only on DF; however it has already been noted that many developers use the final feasibility study as the project monitoring tool as the project progresses, recording actual expenditure and income against the projected, and the fact that Estate Master DF integrates so closely with DM is a boon.

As we will see over the next few pages, DF provides the user with an experience that is similar to that gained from using a well-developed Excel spreadsheet; indeed this is perhaps deliberate. Someone who was used to self-created Excel models would find the transition to DF far easier than perhaps the transition to Argus Developer (see the previous chapter). Both DF and Developer are, however, both quite intuitive to use.

This Microsoft Excel user familiarity philosophy can be seen from the initial screen. Like Excel, DF has a ribbon toolbar and a 'backstage' button that allows some administration to take place. The appraisal is also constructed using a series of Excel-like worksheets which can be observed in the bottom part of the screen in Figure 13.1.

In terms of data entry from the initial screen, Estate Master DF only requires a project name and a unique property reference number to work, although the remaining information will be useful for identification and for later reporting.

Figure 13.1 Estate Master DF opening screen

Before commencing the appraisal data entry it is sensible (and sometimes necessary) to review the Preferences screen (linked via a button on the ribbon toolbar). The preference screen is displayed as Figure 13.2.

The preference screen allows the program to be set up to suit the country details (currency, VAT/GST tax regimes, stamp duty etc.). The program allows considerable flexibility in set up, particularly as regards financing regimes. Multiple loans are catered for in a variety of different types, or else a simple debt/equity calculation mode can be selected. Joint ventures can be modelled, with a specific developer/land owner JV module.

The case of developer/developer JVs is handled by provision within the financing set-up, which also allows for lending to be with participation, i.e. equity sharing. The taxation set-up is only for VAT/GST and no other forms of taxation but that is to be expected; the taxation regimes of individual developers operating in different countries is very complex and it would be unlikely that a single software program could be sufficiently flexible to allow calculation to occur in every regime.

Once the development preferences have been set up, the user can then switch to the first worksheet, the development inputs (Figure 13.3). The first input is entitled preliminary and is used primarily for reporting. It gives the broad outline of the scheme being appraised, its type, the broad gross and net areas and the site ratio (built area to undeveloped). This latter ration is often used in Australia in planning consents/zoning of sites and is thus a key ratio to monitor. It should be noted that none of these figures is actually used in the calculations.

Once the preliminary details have been noted, the next input area is that concerned with land purchase and any of the costs associated with this

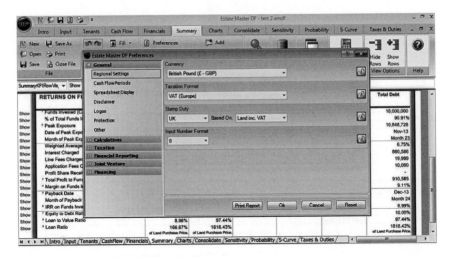

Figure 13.2 Preference screen

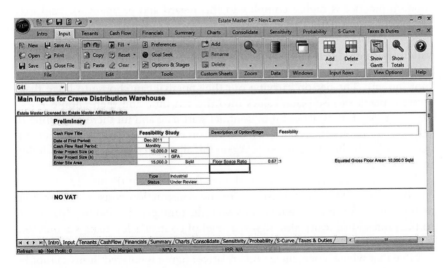

Figure 13.3 Main inputs – preliminary

(Figures 13.4 and 13.5). The land cost may be paid in full or else stage payments are allowed.

The stage payments and/or the land purchase can be modelled to occur at any point during the project. This flexibility is very important as although many development projects do start in the classical way with the acquisition of land,

Main Inputs for Crewe Distribution Warehouse

Estate Master Licensed to: Estate Master Affiliates/Mentors

Land Purchase Price	500,000

Code	Stage		% of Land Purchase Price		AND/OR
			% paid	Amount	Lump Amount
1002	-	Deposit in Trust Account[1]	0.00%	-	-
1003	-	Payment 1	0.00%	-	-
1004	-	Payment 2	0.00%	-	-
1005	-	Payment 3	0.00%	-	-
1006	-	Payment 4	0.00%	-	-
1007	-	Settlement (Balance)	100.00%		500,000
1008	-	Stamp Duty[1]	UK		24,000
		Interest on Deposit in Trust Account	0.00%	Interest from deposit shared between parties	
		Profit Share to Land Owner	0.00%	Paid progressively as project makes a profit.	

Code	Stage	Other Acquisition Costs (to be entered Exclusive of VAT)	% of Land Price exc Tax		AND/OR
			% paid	Amount	Lump Amount
1011	-	Survey	0.00%	-	5,000
1012	-	Acquisition Agent	1.00%	5,000	-
1013	-	Acquisition Legal	1.00%	5,000	-
1014	-		0.00%	-	-
1015	-		0.00%	-	-

[1](No VAT credit available for Stamp Duty)

Figure 13.4 Land purchase

modern practice is highly complex in this regard and payment for land may be deferred until quite late in a project.

Note that the system does have the provision to calculate a residual land value, although this is left to a function within the summary results area.

There are sections in the program which allow for cost and value escalation (not shown). The person carrying out the appraisal can choose to leave these sections blank or else place values into the schedule and then choose N for no escalation; E is escalates up to contract start date and nil thereafter while R simulates a rise and fall contract in costs, something that is now rare but still can occur in the building industry. Again this allows a good degree of flexibility in modelling projects.

The next input screen is that for the professional team (Figure 13.6). These can be either entered as percentages of the construction costs or else entered as lump sums. Again there is flexibility as to the timing and duration of payments to the professional team, although the level of control is less than is achievable with either Excel or Developer (see below).

Note that where a percentage (i.e. relative) figure is entered, sums cannot be calculated until the next section, construction costs, is completed (Figure 13.7).

In keeping with the rest of the program, a great deal of flexibility is displayed in the construction cost calculation screen (although it is still less flexible and comprehensive than the Developer construction cost calculation screen which was reviewed earlier). Construction costs can be entered as a full, elemental breakdown, on a building-by-building basis, or as a simple overall global cost estimate, as is the case in my example here. Again there is considerable control possible in regards to timing and duration of expenditure. A number of s-curve models exist within the system or the user can create their own, custom distribution.

Add VAT on Land Price? Y
Reclaim All After Final Land Settlement

Month Start	Month Span	Cash Flow Period		Total Current Costs (exc. VAT)	Total Current Costs (inc. VAT)	Total Escalated Cost
0	–	–		–	–	–
0	–	–		–	–	–
0	–	–		–	–	–
0	–	–		–	–	–
1	1	Jan-12 – Jan-12		500,000	600,000	600,000
1	1	Jan-12 – Jan-12	Stamp Duty	24,000	24,000	24,000
			TOTAL	524,000	624,000	624,000

(Stamp Duty calculated on Land Value of 600,000 inc. VAT)

Month Start	Month Span	Cash Flow Period	Add VAT	Total Current Costs (exc. VAT)	Total Current Costs (inc. VAT)	Total Escalated Cost	Remarks
2	1	Feb-12 – Feb-12	Y	5,000	6,000	6,000	
1	1	Jan-12 – Jan-12	Y	5,000	6,000	6,000	
1	1	Jan-12 – Jan-12	Y	5,000	6,000	6,000	
			Y	–	–	–	
0	–	–		–	–	–	Manual Input (refer to Cash Flow)
			TOTAL	15,000	18,000	18,000	

and Payments ('L')

Figure 13.5 Land purchase – right-hand side of screen

Main Inputs for Crewe Distribution Warehouse

Estate Master Licensed to: Estate Master Affiliates/Mentors

Code	Stage	Description	% of Construct.[1]	AND/OR No. Units	Base Rate/Unit	Escalate (E,R,N)	S-Curve	Month Start[2]	Month Span	Cash Flow Period
3001	–	Architect	4.00%	–		n	–	6	12	Jun-12 – May-13
3002	–	Quantity Surveyor	2.00%	–		n	–	6	12	Jun-12 – May-13
3003	–	Structural Engineer	1.00%	–		n	–	6	12	Jun-12 – May-13
3004	–	M&E Engineer	1.00%	–		n	–	6	12	Jun-12 – May-13
3005	–	Site Safety	0.50%	–		n	–	6	12	Jun-12 – May-13
3006	–		0.00%	–		–	–	0	–	–
3007	–		0.00%	–		–	–	0	–	–
3008	–		0.00%	–		–	–	0	–	–
3009	–		0.00%	–		–	–	0	–	–
3010	–		0.00%	–		–	–	0	–	–
3011	–		0.00%	–		–	–	0	–	–
3012	–		0.00%	–		–	–	0	–	–
3013	–		0.00%	–		–	–	0	–	–
3014	–		0.00%	–		–	–	0	–	–
3015	–		0.00%	–		–	–	0	–	–

[1] % Based on Net Costs

| 3099 | – | Development Management | 2.00% | % of Project Costs (inc land) excludes finance costs and tax (if applicable) | | – | – | 1 | 24 | Jun-12 – Dec-13 |

[2] Pro-rata with Construction ('C')

[2] Pro-rata with Construction ('C'), Settlements ('S') Project Costs inc Land ('P1') or exc Land ('P2')

Figure 13.6 Modelling professional fees associated with the development

Version 5.13 August 2011

Add VAT	Remarks	Total Current Costs (exc VAT)	Total Current Costs (inc VAT)	Total Escalated Cost
Y		147,000	176,400	176,400
Y		73,500	88,200	88,200
Y		36,750	44,100	44,100
Y		36,750	44,100	44,100
Y		18,375	22,050	22,050
Y		–	–	–
Y		–	–	–
Y		–	–	–
Y		–	–	–
Y		–	–	–
Y		–	–	–
Y		–	–	–
Y		–	–	–
Y		–	–	–
Y		–	–	–
Y		100,710	120,852	120,852
	Manual Input (refer ro Cash Flow)	–	–	–
	TOTAL	413,085	495,702	495,702

Version 5.13 August 2011

Add VAT	Remarks	Total Current Costs (exc VAT)	Total Current Costs (inc VAT)	Total Escalated Cost
Y		3,500,000	4,200,000	4,200,000
Y		–	–	–
Y		–	–	–
Y		–	–	–
Y		–	–	–
Y		–	–	–
Y		–	–	–
Y		–	–	–
Y		–	–	–
Y		–	–	–
Y		–	–	–
Y		–	–	–
Y		–	–	–
Y		–	–	–
Y		–	–	–
Y		–	–	–
Y		–	–	–
Y		–	–	–
	Manual Input (refer ro Cash Flow)	–	–	–
	Construction Contingency	175,000	210,000	210,000
	TOTAL	3,675,000	4,410,000	4,410,000

Figure 13.7 Construction costs calculation sheet

Other costs that can be modelled include statutory fees (Figure 13.8), which in the UK would include such items as Section 106 planning agreements with the Local Authority, payments to the planning authorities and also commuted payments to the Highways authority for road alterations. There is also provision in the system for the calculation of letting and sale fees (not shown).

Later models will illustrate how residential developments are modelled. In this case we are dealing with a commercial, income producing investment type property that will be valued using the income approach. In DF the income stream and the capital value is modelled in a separate worksheet, the Tenancy Schedule (Figure 13.9). Again a high degree of flexibility is possible, multiple buildings or parts of buildings can be appraised separately and assumptions made

Statutory Fees

5000

Costs to be entered Exclusive of VAT

Code	Stage	Description	Units	Base Rate/Units	Escalate (E,R,N)	S-Curve	Month Start	Month Span	Cash Flow Period	Add VAT	Remarks	Total Current Costs (exc VAT)	Total Current Costs (inc VAT)	Total Escalated Cost
5001	–	Section 102 agreement	1	150,000	n	–	18	1	Jun-13 – Jun-13	Y		150,000	180,000	180,000
5002	–	Planning	1	15,000	n	–	3	1	Mar-12 – Mar-12	Y		15,000	18,000	18,000
5003	–	Highways	1	100,000	n	–	16	1	Apr-13 – Apr-13	Y		100,000	120,000	120,000
5004	–		–	–	–	–	0	–	–	Y		–	–	–
5005	–		–	–	–	–	0	–	–	Y		–	–	–
5006	–		–	–	–	–	0	–	–	Y		–	–	–
5007	–		–	–	–	–	0	–	–	Y		–	–	–
5008	–		–	–	–	–	0	–	–	Y		–	–	–
5009	–		–	–	–	–	0	–	–	Y		–	–	–
5010	–		–	–	–	–	0	–	–	Y		–	–	–
5011	–		–	–	–	–	0	–	–	Y		–	–	–
5012	–		–	–	–	–	0	–	–	Y		–	–	–
5013	–		–	–	–	–	0	–	–	Y		–	–	–
5014	–		–	–	–	–	0	–	–	Y		–	–	–
5015	–		–	–	–	–	0	–	–	Y		–	–	–
											Manual Input (refer ro Cash Flow)	–		
											TOTAL	265,000	318,000	318,000

Figure 13.8 Statutory fees

Tenancy Schedule

| 12000 | | Rental Income & Capitalised Sales | | | | | | | | | |

Code	Stage	Description	Total Area SqM	Current Rent /SqM/annum	Outgoings and Vacancies Amount /SqM/annum	Outgoings and Vacancies % of Rent	Outgoings and Vacancies Total Per Annum	Pre-Commit Month	Lease Month Start	Lease Month Span	Cash Flow Period
12001	–	Distribution Warehouse	10,000	75	–	0.00%	–	12	24	18	Dec-13 – May-13
12002	–		–	–	–	0.00%	–	–	–	–	–
12003	–		–	–	–	0.00%	–	–	–	–	–
12004	–		–	–	–	0.00%	–	–	–	–	–
12005	–		–	–	–	0.00%	–	–	–	–	–
12006	–		–	–	–	0.00%	–	–	–	–	–
12007	–		–	–	–	0.00%	–	–	–	–	–
12008	–		–	–	–	0.00%	–	–	–	–	–
12009	–		–	–	–	0.00%	–	–	–	–	–
12010	–		–	–	–	0.00%	–	–	–	–	–
TOTAL			10,000.00								

VAT is payable on settlement (end of lease period or settlement date
value = annual income (net of outgoings) divided by the capitalisation rate

VAT on Rents	Residual Cap Rate	Pre-Sale Exchange Month	Settlement Month	Letting Void Months Vacant	Discount Rate	Purchaser's Costs	VAT Included on Sales
Y	6.25%	–	–	12	7.00%	6.00%	Y
Y	0.00%	–	–	–	0.00%	0.00%	Y
Y	0.00%	–	–	–	0.00%	0.00%	Y
Y	0.00%	–	–	–	0.00%	0.00%	Y
Y	0.00%	–	–	–	0.00%	0.00%	Y
Y	0.00%	–	–	–	0.00%	0.00%	Y
Y	0.00%	–	–	–	0.00%	0.00%	Y
Y	0.00%	–	–	–	0.00%	0.00%	Y
Y	0.00%	–	–	–	0.00%	0.00%	Y
Y	0.00%	–	–	–	0.00%	0.00%	Y

Current Net Annual Rent	Current End Sale Value	Total Net Rental Income less Incentives	Escalated End Sale Value
750,000	12,000,000	567,000	11,133,216
–	–	–	–
–	–	–	–
–	–	–	–
–	–	–	–
–	–	–	–
–	–	–	–
–	–	–	–
–	–	–	–
–	–	–	–
750,000	12,000,000	567,000	11,133,216

Figure 13.9 Tenancy schedule

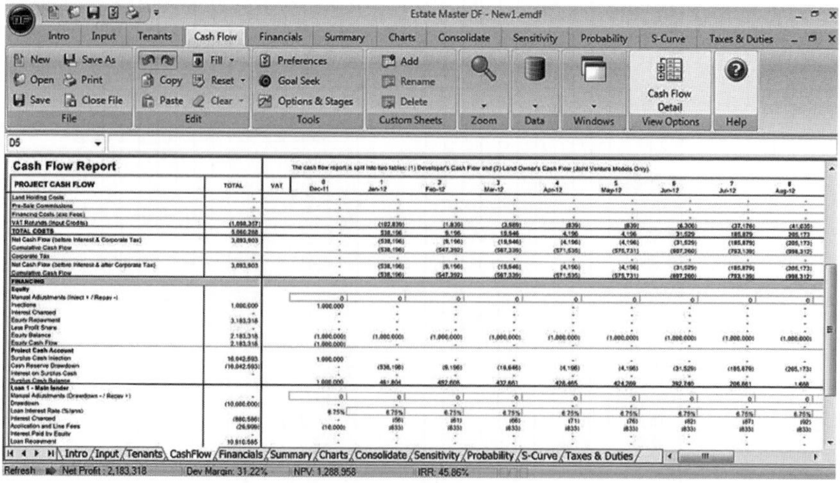

Figure 13.10 Estate Master DF cash flow extract

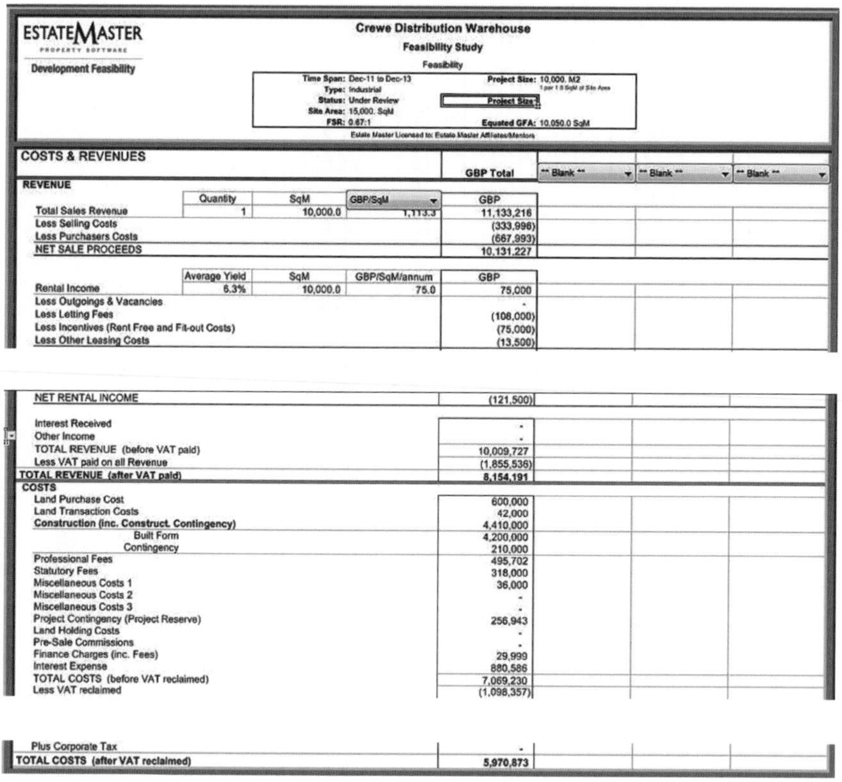

ESTATEMASTER PROPERTY SOFTWARE

Development Feasibility

Crewe Distribution Warehouse

Feasibility Study

Feasibility

Time Span: Dec-11 to Dec-13 | Project Size: 10,000. M2
Type: Industrial
Status: Under Review
Site Area: 15,000. SqM
FSR: 0.67:1 | Equated GFA: 10,050.0 SqM

Estate Master Licensed to: Estate Master Affiliates/Mentors

COSTS & REVENUES				GBP Total	** Blank **	** Blank **	** Blank **
REVENUE							
	Quantity	SqM	GBP/SqM	GBP			
Total Sales Revenue	1	10,000.0	1,113.3	11,133,216			
Less Selling Costs				(333,996)			
Less Purchasers Costs				(667,993)			
NET SALE PROCEEDS				10,131,227			
	Average Yield	SqM	GBP/SqM/annum	GBP			
Rental Income	6.3%	10,000.0	75.0	75,000			
Less Outgoings & Vacancies				-			
Less Letting Fees				(108,000)			
Less Incentives (Rent Free and Fit-out Costs)				(75,000)			
Less Other Leasing Costs				(13,500)			
NET RENTAL INCOME				(121,500)			
Interest Received				-			
Other Income				-			
TOTAL REVENUE (before VAT paid)				10,009,727			
Less VAT paid on all Revenue				(1,855,536)			
TOTAL REVENUE (after VAT paid)				8,154,191			
COSTS							
Land Purchase Cost				600,000			
Land Transaction Costs				42,000			
Construction (inc. Construct. Contingency)				4,410,000			
Built Form				4,200,000			
Contingency				210,000			
Professional Fees				495,702			
Statutory Fees				318,000			
Miscellaneous Costs 1				36,000			
Miscellaneous Costs 2				-			
Miscellaneous Costs 3				-			
Project Contingency (Project Reserve)				256,943			
Land Holding Costs				-			
Pre-Sale Commissions				-			
Finance Charges (inc. Fees)				29,999			
Interest Expense				880,586			
TOTAL COSTS (before VAT reclaimed)				7,069,230			
Less VAT reclaimed				(1,098,357)			
Plus Corporate Tax				-			
TOTAL COSTS (after VAT reclaimed)				5,970,873			

Figure 13.11 Summary screen

about rental values, date of letting, terms, rent free periods and factors affecting capital values such as yield (discount rate) and incoming purchaser's costs can be made. Escalation allowances can also be made. The resultant income and values are then transferred to the cash flow along with all the other cost/value assumptions made previously (Figure 13.10).

The cash flow is locked for editing but adjustment can be made directly into it (like both Excel and Argus Developer) and a direct transfer of the data into Excel is possible.

The outcome of the development appraisal can then be viewed and printed via the summary sheet (Figure 13.11). There is also a comprehensive sensitivity analysis module which allows the robustness of the outcome to be examined (not shown).

The summary sheet also contains the performance indicators (Figure 13.12) and also a very useful financial analysis section (Figure 13.13). Both of these screens provide the user/developer with extremely useful data, better in many respects than that obtainable with the standard module of Argus Developer (see below). This, when placed alongside the program's transparency and ease of use makes it a very valuable tool for the developer and development advisor. It has all the qualities of a really good self-created Excel model (as noted in the introduction) but has few of the downside risks that come with using Excel – the risks of error, auditing problems and alterations discussed earlier.

Estate Master DF case study one – profitability calculation

It is instructive to model the same development in Estate Master DF as has just been completed for Argus Developer, not to play one off against the other but rather to explore the differences and/or similarities in approach between the two systems.

One important thing to stress is that both use the same calculation approach; both are pure cash flow models, using the residual layout only as an output for the development results. The primary difference is how the development data is input. Developer makes extensive use of dedicated input screens and tabs with the calculation detail a few layers back. You can drill down into the detail if required but more of the calculation is done deeper in the program. With DF, the input is closer to the calculation, effectively sitting only one layer above.

Data entry in DF is via a series of Excel-like worksheets, again underlying the fact that, philosophically, the software is closer to the Excel way of working than Developer is.

The project input screen (Figure 13.14) is where the project is defined and identified, and where the project file is created and saved. It is also sensible to work on the project preferences at this point where the currency, stamp duty and tax (VAT) regime is defined.

PERFORMANCE INDICATORS

[1] **Net Development Profit**	**2,183,318**	
[3] **Development Margin (or Profit/Risk Margin)**	**31.22%**	on total development costs (inc selling costs).
[4] Residual Land Value (based on 20% Target Margin)	N.A.	Unable to Compute
[5] **Net Present Value**	**1,288,958**	(at 20% per ann. discount rate, nominal)
[6] Benefit Cost Ratio	1.3073	(at 20% per ann. discount rate, nominal)
[7] **Project Internal Rate of Return**	**45.86%**	(per ann. nominal)
[8] Residual Land Value (based on NPV)	1,706,914	(Exclusive of VAT) 113.79 per SqM of Site Area
Equity IRR	59.32%	(per ann. nominal)
Equity Contribution	1,000,000	
Peak Debt Exposure	10,848,728	
Equity to Debt Ratio	10.00%	
[9] Weighted Average Cost of Capital (WACC)	6.14%	
[10] Breakeven Date for Cumulative Cash Flow	Dec-2013	(Month 24)
[11] Yield on Cost	9.29%	
[12] Rent Cover	2 Yrs, 11 Mths	
[13] Profit Erosion	2 Yrs, 11 Mths	

Figure 13.12 Summary screen – performance indicators

RETURNS ON FUNDS INVESTED	Equity	Loan 1 Main lender			Total Debt
1 Funds Invested (Cash Outlay)	1,000,000	1,000,000			1,000,000
% of Total Funds Invested	9.09%	90.91%			90.91%
2 Peak Exposure	1,000,000	10,848,728			10,848,728
Date of Peak Exposure	Dec-11	Nov-13			Nov-13
Month of Peak Exposure	Month 0	Month 23			Month 23
Weighted Average Interest Rate	N.A.	6.75%			6.75%
Interest Charged	–	880,586			880,586
Line Fees Charged	–	19,999			19,999
Application Fees Charged	–	10,000			10,000
Profit Share Received	–	–			–
3 Total Profit to Funders	2,183,318	910,585			910,585
4 Margin on Funds Invested	218.33%	9.11%			9.11%
5 Payback Date	Dec-13	Dec-13			Dec-13
Month of Payback	Month 24	Month 24			Month 24
6 IRR on Funds Invested	59.32%	6.99%			6.99%
7 Equity to Debt Ratio		10.00%			10.00%
8 Loan to Value Ratio	8.98%	97.44%			97.44%
9 Loan Ratio	166.67%	1818.43%			1818.43%
	of Land Purchase Price	of Land Purchase Price			of Land Purchase Price

Figure 13.13 Summary screen – financial analysis

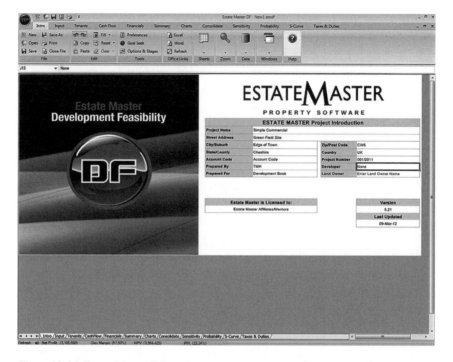

Figure 13.14 Estate Master DF project input screen – simple commercial project

Primary data entry is, logically, via the Input worksheet. The Input worksheet is divided into a series of functional areas that are accessed by scrolling down the page.

The top part of the screen is the preliminary section (Figure 13.15). The data entered here is only used in reporting and has no impact on the calculation itself.

The next section deals, the first effective input screen, with land acquisition. As we are calculating the developments projected profitability, the land purchase price should be known and can be entered. If the programme is being run in land residual mode, either a nominal sum is entered or this box is left blank. A residual calculation at a defined target profit goal figure can be run on the summary worksheet when all the other assumptions have been defined and entered.

Here the land purchase is priced at £2,000,000. The stamp duty is automatically calculated but the associated costs (acquisition agent, legal fees, town planning fees and survey) have been manually entered and then the distribution defined in the columns to the right (Figure 13.16). Note that staged payments for land or specific payments on dates can be defined in the program.

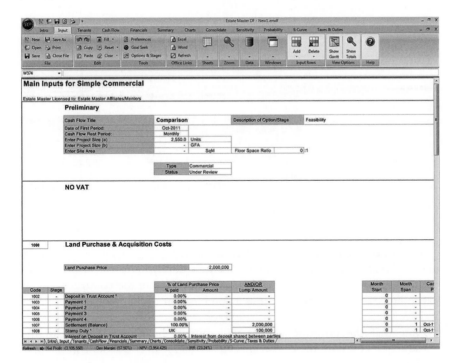

Figure 13.15 DF input screen – top

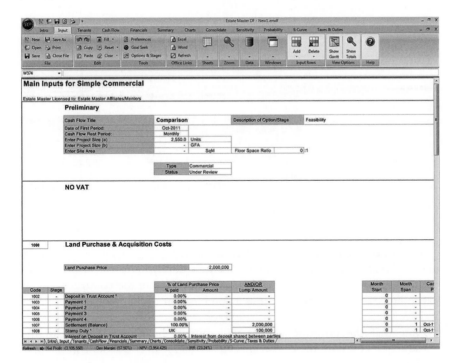

Figure 13.16 Land purchase and acquisition costs

The totals are found on the right-hand side of the screen (Figure 13.17).

There is substantial allowance for defining escalation rates in costs and values in DF. Whilst this is essential in some markets (see the section in the introduction) it does make the appraisal far more volatile and should be avoided if possible. This appraisal (as with the Argus Developer version of the calculation) has been carried out flat, i.e. without any inflation or growth assumptions.

	Total Current Costs (exc VAT)	Total Current Costs (inc VAT)	Total Escalated Cost
	-	-	-
	-	-	-
	-	-	-
	-	-	-
	-	-	-
	2,000,000	2,000,000	2,000,000
(Stamp Duty calculated on Land Value of 2,000,000 exc. VAT) **Stamp Duty**	100,000	100,000	100,000
TOTAL	2,100,000	2,100,000	2,100,000

Remarks	Total Current Costs (exc VAT)	Total Current Costs (inc VAT)	Total Escalated Cost
	20,000	20,000	20,000
	20,000	20,000	20,000
	10,000	10,000	10,000
	5,000	5,000	5,000
	-	-	-
Manual Input (refer to Cash Flow)	-	-	-
TOTAL	55,000	55,000	55,000

Figure 13.17 The totals of the land acquisition and the associated costs

The next section after the escalation deals with professional fees. It can be observed that there is considerable scope for flexibility in the calculations, with the option of defining the cost as being a percentage of the construction cost expenditure or as a lump sum, or as a combination of the two (Figures 13.18 and 13.19).

Note that the professional fees will initially be zero if related to the construction cost until these costs are actually defined. This is done in the next section.

This project is very simple, a single building with associated external works. Estate Master DF allows for any number of buildings and building types and can model the construction expenditure using pre-set curves (e.g. the classic s-curve) or by any user defined definition. Here one of the pre-set s-curves has been used (Figures 13.20 and 13.21).

One of the miscellaneous cost data entry points has been used to model the expenditure on marketing (Figure 13.22). The assumption made is clear; the budgeted lump sum, with expenditure starting in month 6 and continuing over 3 months with flat, rather than s-curved expenditure.

Where Estate Master DF differs markedly from the base version of Argus Developer is in regards to financing. Whilst the base version of the latter has a relatively simple finance module, DF is able to separately analyse debt and equity sources of funding (Figure 13.23).

The issues of opportunity costs and finance rates has already been discussed in Part One; the argument generally used is that the opportunity cost of using equity funds in a development should be similar to that applied to lending on the same project, and, therefore 100 per cent debt financing tends to be assumed in development feasibility studies as a simplifying step.

In reality, however, developers DO often seek different returns from different sources of funds and, perhaps more importantly, one of the key measures of project return to the developer is not the overall project return but the return on equity funds employed.

Professional Fees

Stage	Description	% of Construct.[1]	AND/OR No. Units	Base Rate/Unit	Escalate (E,R,N)	S-Curve	Month Start[2]	Month Span	Cash Flow Period
-	Architect	4.00%	-	-	-	-	3	6	Jan-12 – Jun-12
-	Quantity Surveyor	2.00%	-	-	-	-	3	6	Jan-12 – Jun-12
-	Structural Engineer	1.00%	-	-	-	-	3	6	Jan-12 – Jun-12
-	M&E Engineer	1.00%	-	-	-	-	3	6	Jan-12 – Jun-12
-	Project Manager	1.00%	-	-	-	-	3	6	Jan-12 – Jun-12
-	Site Safety	0.50%	-	-	-	-	3	6	Jan-12 – Jun-12
-		0.00%	-		-	-	0	-	-
-		0.00%	-		-	-	0	-	-
-		0.00%	-		-	-	0	-	-
-		0.00%	-		-	-	0	-	-
-		0.00%	-		-	-	0	-	-
-		0.00%	-		-	-	0	-	-
-		0.00%	-		-	-	0	-	-

[1] % Based on Net Costs

| - | Development Management | 0.00% | % of Project Costs (inc land) excludes finance costs and tax (if applicable) | | | - | 0 | 0 | - |

[2] Pro-rata with Construction ('C')

[2] Pro-rata with Construction ('C'), Settlements ('S')
Project Costs inc Land ('P1') or exc Land ('P2')

Figure 13.18 Definition and distribution of professional fees

Remarks	Total Current Costs (exc VAT)	Total Current Costs (inc VAT)	Total Escalated Cost
	112,000	112,000	112,000
	56,000	56,000	56,000
	28,000	28,000	28,000
	28,000	28,000	28,000
	28,000	28,000	28,000
	14,000	14,000	14,000
	–	–	–
	–	–	–
	–	–	–
	–	–	–
	–	–	–
	–	–	–
	–	–	–
	–	–	–
	–	–	–
	–	–	–
Manual Input (refer ro Cash Flow)	–	–	–
TOTAL	266,000	266,000	266,000

Figure 13.19 The totalled sums of the professional fees

4000 **Construction Costs**

Code	Stage	Description	Cost Type	Units	Base Rate / Units	Escalate (E,R,N)1	S-Curve	Month Start	Month Span	Cash Flow Period
4001	-	Main Office build	BUI	3,000	900	n	S	3	6	Jan-12 - Jun-12
4002	-	Road and Site Works	BUI	1	100,000	n	s	5	3	Mar-12 - May-12
4003	-	,	-	-	-	-	-	0	-	-
4004	-	,	-	-	-	-	-	0	-	-
4005	-	,	-	-	-	-	-	0	-	-
4006	-	,	-	-	-	-	-	0	-	-
4007	-	,	-	-	-	-	-	0	-	-
4008	-	,	-	-	-	-	-	0	-	-
4009	-	,	-	-	-	-	-	0	-	-
4010	-	,	-	-	-	-	-	0	-	-
4011	-	,	-	-	-	-	-	0	-	-
4012	-	,	-	-	-	-	-	0	-	-
4013	-	,	-	-	-	-	-	0	-	-
4014	-	,	-	-	-	-	-	0	-	-
4015	-	,	-	-	-	-	-	0	-	-
4016	-	,	-	-	-	-	-	0	-	-
4017	-	,	-	-	-	-	-	0	-	-
4018	-	,	-	-	-	-	-	0	-	-
4019	-	,	-	-	-	-	-	0	-	-
4025	-	,	-	-	-	-	-	0	-	-

[1] Escalation ('N' = no escalation, 'E' = escalation to start period, 'R' = escalation to start period and through span)

4099 **Construction Contingency** - **And / Or** 5.00% of Construction Costs (inc VAT)

Figure 13.20 Construction costs entry and distribution

Remarks	Total Current Costs (exc VAT)	Total Current Costs (inc VAT)	Total Escalated Cost
	2,700,000	2,700,000	2,700,000
	100,000	100,000	100,000
	–	–	–
	–	–	–
	–	–	–
	–	–	–
	–	–	–
	–	–	–
	–	–	–
	–	–	–
	–	–	–
	–	–	–
	–	–	–
	–	–	–
	–	–	–
	–	–	–
	–	–	–
	–	–	–
Manual Input (refer ro Cash Flow)	–	–	–
Construction Contingency	140,000	140,000	140,000
TOTAL	2,940,000	2,940,000	2,940,000

Figure 13.21 Construction costs entry and distribution totals

6000

Miscellaneous Costs 1

Code	Stage	Description	% of Construction[1]	AND/OR No. Units	Base Rate/Unit	Escalate (E,R,N)	S-Curve	Month Start[2]	Month Span	Cash Flow Period
6001	–	Marketing	0.00%	1	25,000	–	–	6	3	Apr-12 – Jun-12
6002	–		0.00%	–	–	–	–	0	–	–
6003	–		0.00%	–	–	–	–	0	–	–
6004	–		0.00%	–	–	–	–	0	–	–
6005	–		0.00%	–	–	–	–	0	–	–
6006	–		0.00%	–	–	–	–	0	–	–
6007	–		0.00%	–	–	–	–	0	–	–
6008	–		0.00%	–	–	–	–	0	–	–
6009	–		0.00%	–	–	–	–	0	–	–
6010	–		0.00%	–	–	–	–	0	–	–

[1] % Based on Net Costs

[2] Pro-rata with Construction ('C'), Settlements ('S')

Figure 13.22 Miscellaneous costs – marketing

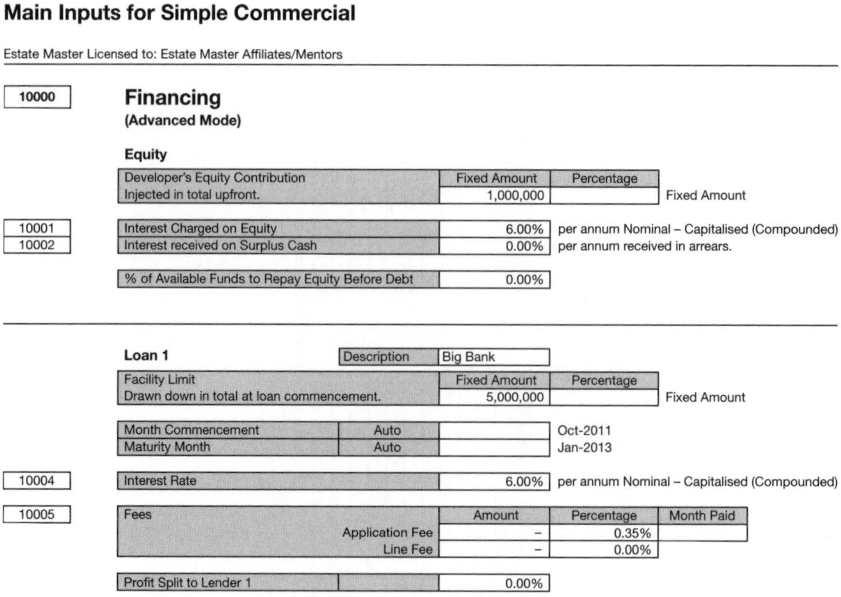

Main Inputs for Simple Commercial
Estate Master Licensed to: Estate Master Affiliates/Mentors
10000
Financing
(Advanced Mode)
Equity
Developer's Equity Contribution
Injected in total upfront.
Fixed Amount
Percentage
1,000,000
Fixed Amount
10001
10002
Interest Charged on Equity
6.00%
per annum Nominal – Capitalised (Compounded)
Interest received on Surplus Cash
0.00%
per annum received in arrears.
% of Available Funds to Repay Equity Before Debt
0.00%
Loan 1
Description
Big Bank
Facility Limit
Drawn down in total at loan commencement.
Fixed Amount
Percentage
5,000,000
Fixed Amount
Month Commencement
Auto
Oct-2011
Maturity Month
Auto
Jan-2013
10004
Interest Rate
6.00%
per annum Nominal – Capitalised (Compounded)
10005
Fees
Amount
Percentage
Month Paid
Application Fee
-
0.35%
Line Fee
-
0.00%
Profit Split to Lender 1
0.00%

Figure 13.23 Financing assumptions

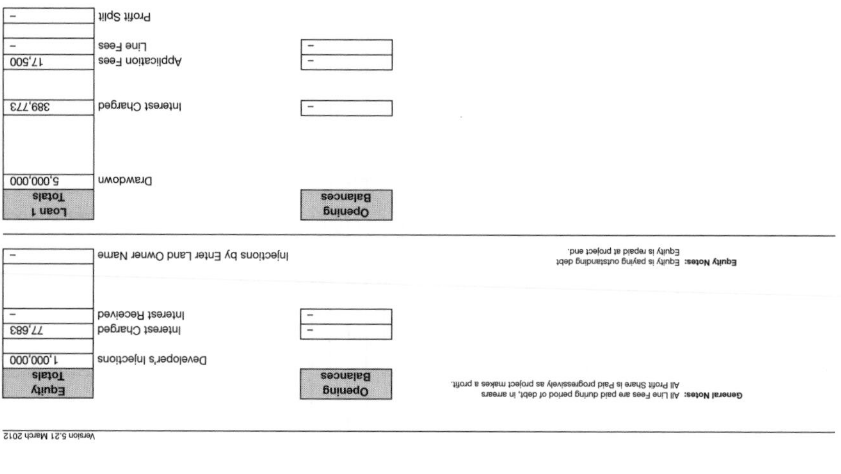

Loan 1 Totals
Drawdown
5,000,000
Interest Charged
389,773
Application Fees
17,500
Line Fees
-
Profit Split
-
Opening Balances
Equity Totals
Developer's Injections
1,000,000
Interest Charged
77,683
Interest Received
-
Injections by Enter Land Owner Name
-
General Notes: All Line Fees are paid during period of debt, in arrears
All Profit Share is Paid progressively as project makes a profit.
Equity Notes: Equity is paying outstanding debt
Equity is repaid at project end.
Version 5.21 March 2012

Figure 13.24 Finance screen totals

It is the ability for DF to produce this returns in its base version that give the program distinct advantages over Developer (Figure 13.24).

Estate Master DF deals with the inputs required for residential development within the Input worksheet, however any commercial element has its own data entry point, the Tenancy Schedule, a separate worksheet. This is probably due to the complexity of the commercial lease and valuation structure.

The details of the commercial elements are entered into a rather complex, lateral worksheet (Figures 13.25 to 13.27) which allows lease start dates, duration, rental value, incentives, escalation rates to be entered as well as the Settlement month (disposal date) and capitalisation rate. This then calculates the end value.

Once all the data is entered the cash flow and the summary sheet, both have which have been constructed automatically, can be inspected (Figures 13.28 and 13.29).

Estate Master DF is particularly strong on performance measures (Figure 13.30) and also on an area noted above, being able to analyse the returns on equity and debt sources of funds (Figure 13.31).

Estate Master DF case study two – sensitivity analysis

This case study uses the data from case study one.

As noted in the introduction, Estate Master DF also has a comprehensive sensitivity analysis module that allows for variables to be tested for simple sensitivity, scenarios and a probability-based scenario.

The simple probability analysis is user defined and allows a snapshot of the impact of changing key variables across the full range of development return

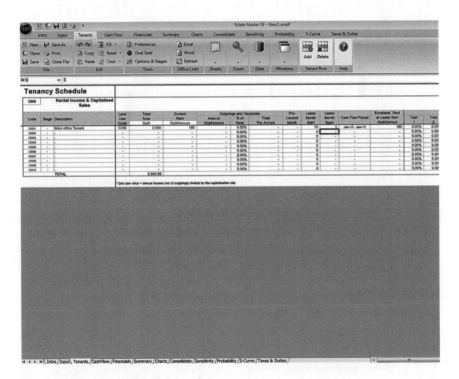

Figure 13.25 Tenancy schedule worksheet

Tenancy Schedule

12000		Rental Income & Capitalised Sales					

Code	Stage	Description	Letting Fee		Rent Free Months	Incentives	
			% paid at PreCommit	Total Amount		Fit out Cost	Month Start
12001	-	Main office Tenant	0.00%	45,900	6	-	15
12002	-		0.00%	-	-	-	-
12003	-		0.00%	-	-	-	-
12004	-		0.00%	-	-	-	-
12005	-		0.00%	-	-	-	-
12006	-		0.00%	-	-	-	-
12007	-		0.00%	-	-	-	-
12008	-		0.00%	-	-	-	-
12009	-		0.00%	-	-	-	-
12010	-		0.00%	-	-	-	-
		TOTAL					

Residual Cap. Rate	Pre-Sale Exchange Month	Settlement Month	Letting Void		Purchaser's Costs
			Months Vacant	Discount Rate	
6.00%	15	15	6	6.00%	6.00%
0.00%	-	-	-	0.00%	0.00%
0.00%	-	-	-	0.00%	0.00%
0.00%	-	-	-	0.00%	0.00%
0.00%	-	-	-	0.00%	0.00%
0.00%	-	-	-	0.00%	0.00%
0.00%	-	-	-	0.00%	0.00%
0.00%	-	-	-	0.00%	0.00%
0.00%	-	-	-	0.00%	0.00%
0.00%	-	-	-	0.00%	0.00%

Licensed to: Estate Master Affiliates/Mentors

Figure 13.26 Setting the lease details in the tenancy schedule

Tenancy Schedule

12000		Rental Income & Capitalised Sales						

Code	Stage	Description	Residual Cap. Rate	Pre-Sale Exchange Month	Settlement Month	Letting Void		Purchaser's Costs
						Months Vacant	Discount Rate	
12001	-	Main office Tenant	6.00%	15	15	6	6.00%	6.00%
12002	-		0.00%	-	-	-	0.00%	0.00%
12003	-		0.00%	-	-	-	0.00%	0.00%
12004	-		0.00%	-	-	-	0.00%	0.00%
12005	-		0.00%	-	-	-	0.00%	0.00%
12006	-		0.00%	-	-	-	0.00%	0.00%
12007	-		0.00%	-	-	-	0.00%	0.00%
12008	-		0.00%	-	-	-	0.00%	0.00%
12009	-		0.00%	-	-	-	0.00%	0.00%
12010	-		0.00%	-	-	-	0.00%	0.00%
		TOTAL						

Current Net Annual Rent	Current End Sale Value	Total Net Rental Income less Incentives	Escalated End Sale Value
459,000	7,650,000	(45,900)	7,424,463
-	-	-	-
-	-	-	-
-	-	-	-
-	-	-	-
-	-	-	-
-	-	-	-
-	-	-	-
-	-	-	-
-	-	-	-
459,000	7,650,000	(45,900)	7,424,463

Figure 13.27 Setting the capitalisation rate and disposal value

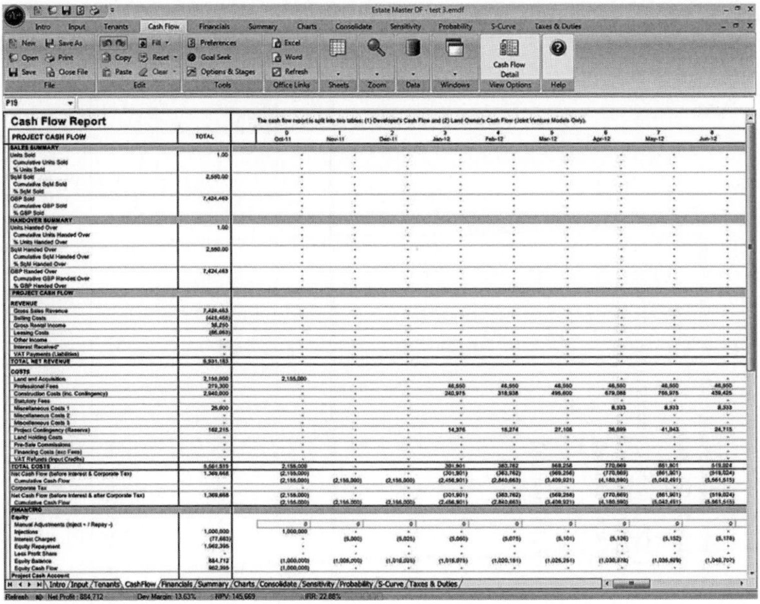

Figure 13.28 DF cash flow extract

Figure 13.29 Summary sheet for simple commercial project

PERFORMANCE INDICATORS

[1] Net Development Profit	**884,712**	
[3] **Development Margin (or Profit/Risk Margin)**	**13.63%**	on total development costs (inc selling costs).
[4] Residual Land Value (based on 20% Target Margin)	1,660,052	Unable to Compute
[5] **Net Present Value**	**145,669**	(at 20% per ann. discount rate, nominal)
[6] Benefit Cost Ratio	1.0277	(at 20% per ann. discount rate, nominal)
[7] **Project Internal Rate of Return**	**22.88%**	(per ann. nominal)
[8] Residual Land Value (based on NPV)	2,135,192	
Equity IRR	55.16%	(per ann. nominal)
Equity Contribution	1,000,000	
Peak Debt Exposure	5,380,371	
Equity to Debt Ratio	20.00%	
[9] Weighted Average Cost of Capital (WACC)	5.00%	
[10] Breakeven Date for Cumulative Cash Flow	Jan-2013	(Month 15)
[11] Yield on Cost	7.07%	
[12] Rent Cover	1 Yrs, 11 Mths	
[13] Profit Erosion	2 Yrs, 3 Mths	

Footnotes:

1. Development Profit: is total revenue less total cost including interest paid and received.
2. Note: No redistribution of Developer's Gross Profit.
3. Development Margin: is profit divided by total development costs (inc selling costs).
4. Residual Land Value: is the maximum purchase price for the land whilst achieving the target development margin.
5. Net Present Value: is the project's cash flow stream discounted to present value. It includes financing costs but excludes interest and corp tax.
6. Benefit Cost Ratio: is teh ration of discounted incomes to discounted costs and includes financing costs but excludes interest and corp tax.
7. Internal Rate of Return: is the discount rate where the NPV above equals zero.
8. Residual Land Value (based on NPV): is the purchase price for the land to achieve a zero NPV.
9. The Weighted Average Cost of Capital (WACC): is the rate that a company is expected to pay to finance its assests.
10. Breakdown date for Cumulative Cash Flow: is the last date when total debt and equity is repaid (i.e. when profit is realised).
11. Yield on Cost: is Current Net Annual Rent divided by Total Costs (before VAT reclaimed) including all selling costs.
12. The total net development profit divided by the current net annual rental expressed as a number of years/months.
13. The period of time post practical completion that it can remain unsold (but leased out) until finance and land holding costs erodes the profit for the development to zero.

Figure 13.30 Performance measures

RETURNS ON FUNDS INVESTED	Equity	Loan 1				Total Debt
		Big Bank				
[1] Funds Invested (Cash Outlay)	1,000,000	5,000,000				5,000,000
% of Total Funds Invested	16.67%	83.33%				83.33%
[2] Peak Exposure	1,072,321	5,380,371				5,380,371
Date of Peak Exposure	Dec-12	Dec-12				Dec-12
Month of Peak Exposure	Month 14	Month 14				Month 14
Weighted Average Interest Rate	6.00%	6.00%				6.00%
Interest Charged	77,683	389,773				389,773
Line Fees Charged	-	-				
Application Fees Charged	-	17,500				17,500
Profit Share Received	-	-				
[3] Total Profit to Funders	962,395	407,273				407,273
[4] Margin on Funds Invested	96.24%	8.15%				8.15%
[5] Payback Date	Jan-13	Jan-13				Jan-13
Month of Payback	Month 15	Month 15				Month 15
[6] IRR on Funds Invested	55.16%	6.28%				6.28%
[7] Equity to Debt Ratio		20.00%				20.00%
[8] Loan to Value Ratio	14.44%	72.47%				72.47%
[9] Loan Ratio	53.86%	270.36%				270.36%
	of Land Purchase Price	of Land Purchase Price				of Land Purchase Price

Footnotes:
1. The total amount of funding injected into the project cash flow.
2. The maximum cash flow exposure of that equity/debt facility including capitalised interest.
3. The total repayments less funds invested, including profit share paid or received.
4. Margin is net profit divided by total funds invested (cash outlay).
5. Payback date for the equity/debt facility is the last date when total equity/debt is repaid.
6. IRR on Funds Invested is the IRR of the equity cash flow including the return of equity and realisation of project profits.
7. Equity to Debt Ratio is the amount of equity contributed into the project as a percentage of debt funding.
8. Loan to Value ratio is the Peak Equity/Debt Exposure divided by Total Sales Revenue.
9. Loan Ratio is the total funds invested by the lender (cash outlay) divided by the nominated ratio calculation method. It includes capitalised interest and fees.

Figure 13.31 Performance measures – return on funds invested

Sensitivity Analysis

ESTATEMASTER
PROPERTY SOFTWARE
Development Feasibility

Simple Commercial
Comparison
Feasibility
Estate Master | licensed to: Estate Master Affiliates/clients

SENSITIVITY TABLE	Change %	Net Dev. Profit	NPV	Dev. Margin	Project IRR	Equity IRR
Base Case (No Variation)	0.0%	884,712	145,669	13.63%	22.88%	55.16%
Land Acquisition Costs	-5.0%	995,075	256,031	15.59%	25.15%	59.75%
	-3.0%	950,963	211,919	14.80%	24.24%	57.95%
	3.0%	818,363	79,319	12.48%	21.55%	52.29%
	5.0%	774,075	35,031	11.72%	20.68%	50.32%
Construction Costs	-10.0%	1,222,739	452,428	19.87%	29.11%	68.53%
	-5.0%	1,053,725	299,049	16.67%	25.97%	62.10%
	5.0%	715,699	(7,711)	10.74%	19.85%	47.65%
	10.0%	546,686	(161,090)	8.00%	16.87%	39.44%
Construction Period [*]	-20.0%	981,022	385,443	15.34%	29.59%	73.60%
	-10.0%	916,976	206,252	14.20%	24.29%	60.42%
	10.0%	819,700	(2,554)	12.50%	19.96%	46.49%
	20.0%	786,951	(82,921)	11.94%	18.63%	42.86%
End Sale Values	-5.0%	525,162	(134,926)	8.12%	17.26%	38.34%
	-3.0%	668,982	(22,688)	10.33%	19.54%	45.46%
	3.0%	1,100,442	314,026	16.92%	26.12%	63.92%
	5.0%	1,244,262	426,264	19.10%	28.22%	69.32%
Capitalisation Rate	-0.5%	1,538,439	655,842	23.55%	32.41%	79.46%
	-0.2%	1,132,678	339,183	17.41%	26.59%	65.16%
	0.2%	652,744	(35,360)	10.08%	19.29%	44.68%
	0.5%	331,558	(286,016)	5.14%	14.10%	27.76%
Sales Span [**]	-30.0%	884,712	145,669	13.63%	22.88%	55.16%
	-20.0%	884,712	145,669	13.63%	22.88%	55.16%
	20.0%	884,712	145,669	13.63%	22.88%	55.16%
	30.0%	884,712	145,669	13.63%	22.88%	55.16%
Rental Levels	-20.0%	(501,525)	(936,160)	-7.83%	-0.64%	-43.31%
	-10.0%	191,594	(395,245)	2.97%	11.76%	19.23%
	10.0%	1,577,830	686,583	24.14%	32.95%	80.73%
	20.0%	2,270,949	1,227,498	34.51%	42.16%	100.69%
Debt Interest Rates	-2.0%	1,017,671	145,669	16.00%	22.88%	60.66%
	-1.0%	951,578	145,669	14.81%	22.88%	57.97%
	1.0%	817,066	145,669	12.46%	22.88%	52.23%
	3.0%	679,397	145,669	10.14%	22.88%	45.95%
Discount Rate	18.0%		250,561			
	19.0%		197,717			
	20.0%		145,669			
	21.0%		94,404			

[*] Variation to Construction Period in sensitivity table delays span dates for Construction and start and span for Professional Fees, Statutory Contributions and Misc. Costs. Delays start but not span date for Sales, Rental and Other Income. Delays the span for Land Holding costs. Has no effect on Land Purchase or Finance costs (except interest).
[**] Varies span date for Pre-Sale Exchange and Settlement periods, but not commencement dates.

Figure 13.32 Simple sensitivity table output

measures (Figure 13.32). Reference to the section in Part One, where the importance of being able to identify where the developer and the development was most vulnerable to changing market conditions and the accuracy of assumptions was stressed, illustrates the advantages of this analytical tool.

Estate Master DF goes further however, with its probability-based scenario analysis module. This is essentially a Monte Carlo simulation module. It enables the user to define variables to be tested and to set the variable range and probability of occurrence for each variable (Figure 13.33). Then, when the simulation is run, predicted outputs for development return and IRR are calculated and charted (Figures 13.34 and 13.35).

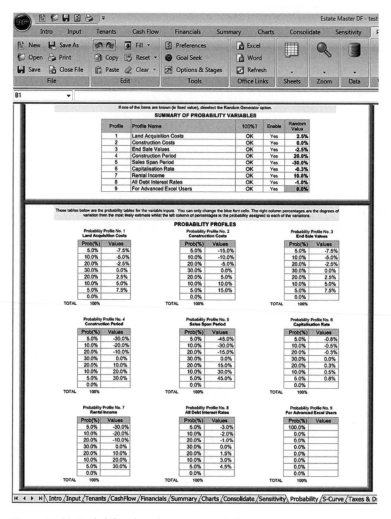

Figure 13.33 Probability-based scenario analysis set-up screen

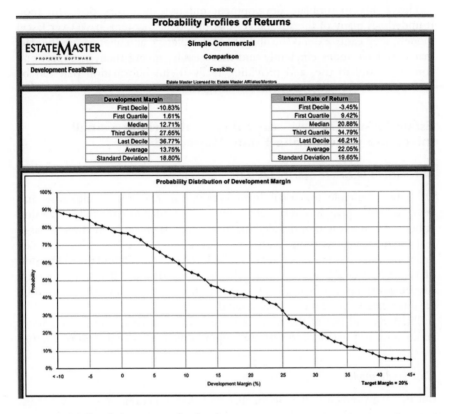

Figure 13.34 Simulation output for development return

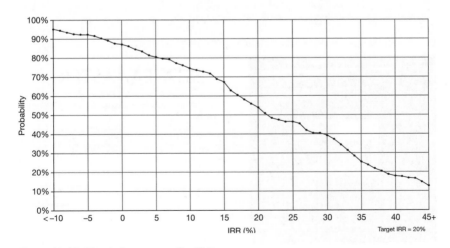

Figure 13.35 Simulation output for IRR

This ability to simulate development outcome is a very powerful tool. It underlines many of the advantages that proprietary systems have over Excel models; they easily extend the abilities of the average appraiser to use the data that they have more efficiently and more easily, giving them a reliable and consistent tool to use, and providing really useful information that can be incorporated into risk management strategies.

Estate Master DF case study three – modelling a small mixed use project using Estate Master DF Lite

Introduction to Estate Master DF Lite

DF Lite is suitable for small firms conducting real estate valuations and development appraisals of relatively simple projects, that find the full DF version either too complicated or detailed for their needs. If you are a medium to large real estate valuer working with large development projects with complex financing, joint venture structures or multiple stages, you will find the full version of Estate Master DF more beneficial; however DF Lite offers some distinct advantages in terms of speed and simplicity that might mean that it would be ideal for valuers providing initial stage development advice.

The look and feel of the new software is very similar to the Estate Master family and, in particular, to DF itself, upon which it is based (Figure 13.36).

Case study model

On the main screen is the ribbon bar with key functions accessed via icons plus tabs, repeated at the bottom of the screen, that access the worksheets.

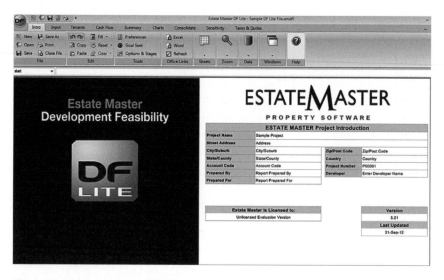

Figure 13.36 Estate Master DF Lite opening screen

Intro is the project identification screen whilst input and tenants are the two primary data and assumption entry sheets that would be used if using the 'full' DF programme (above).

The difference with DF Lite comes when you click on the input tab. This opens up the quick set-up wizard on the ribbon bar (Figure 13.37). Note that all of the worksheets have sheet specific icons that open up when working in these areas.

Clicking on the set-up wizard opens up a series of 15 screens that the user inputs data into. This allows the rapid construction of a development appraisal and avoids the need to enter into the specific worksheets (although once set up the user can refine the development assumptions in exactly the same way as can be done in DF and DM).

The first screen (Figure 13.38) is for project identification. This information is also used in the reporting module of the software.

The second screen provides more information on the project itself (Figure 13.39). Most of the information – on project type, size and the site area – is used in reporting and analysis and does not affect the actual calculation however the project start does. Clicking on the project start box opens up a calendar enabling the project start day to be selected. An indicative completed

Figure 13.37 Ribbon bar

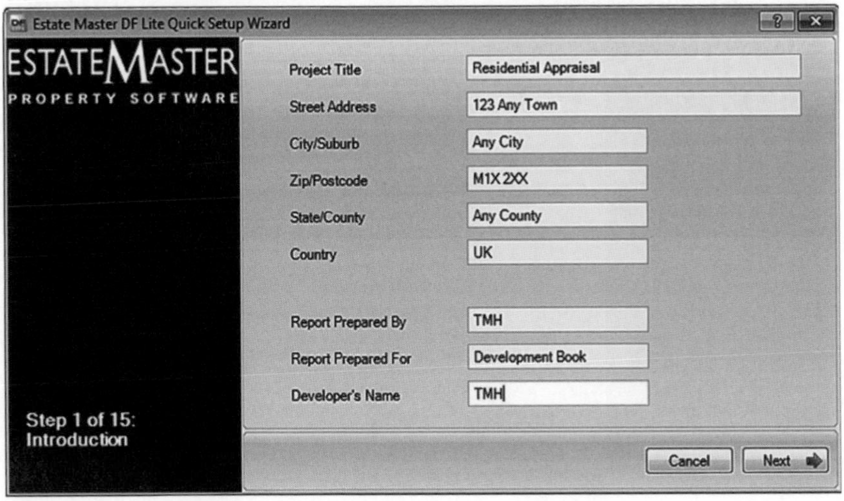

Figure 13.38 Quick set-up screen 1

second screen is shown above. Note that the programme does require a project type and size in order for the user to proceed.

The third screen is where the currency that the appraisal is done in is defined and also what, if any, tax regime applies (Figure 13.40). Note that most of the world's main currencies exist within the system (though there is no provision for exchange rate conversion). The taxation system is for VAT/GST type taxes rather than capital or income taxation.

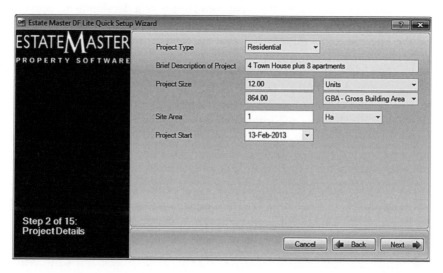

Figure 13.39 Quick set-up screen 2

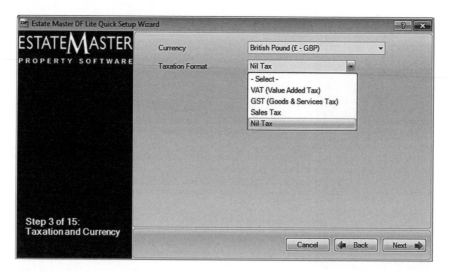

Figure 13.40 Quick set-up screen 3

The fourth screen deals with the detail of the land. If the land purchase price is known it can be entered here as are the assumptions regarding the details of the purchase (Figure 13.41). Estate Master DF Lite has a land residual calculation function if the system is being used for land valuation/land bid calculation but this is carried out as a final step once all the inputs have been made. It is therefore necessary to either enter a nominal sum in the land purchase price box or an approximate figure that will be refined later using the assumptions about cost, values and desired returns.

The fifth screen (Figure 13.42) sets the inflation (costs) and growth (values) assumptions. The 'Full' programme allows much finer control of these assumptions – and of course the assumptions here can be refined once the wizard has been completed – but here single, global figures are used.

The sixth screen in the wizard is where the broad assumptions about the professional team costs are defined (Figure 13.43). Again the level of definition and control is not as great as with the full program but the use of a global cost for the construction professionals is perfectly adequate for the majority of initial appraisals and, once again, the details can be refined post-appraisal completion. The wizard does allow the development management to be defined separately and gives the most common options for its calculation.

Step 7 of the wizard is where the construction costs are entered. In this case as we are doing a residential scheme this can be defined as per unit (in this case apartment) or as a total sum (above). Step 7 also requires you to define the key timing assumptions (start date and duration), s-curve assumptions and contingency percentage (Figure 13.44). Again this is usually perfectly adequate for the majority of simple appraisals but the full program allows much more control and definition at the set-up stage. Such refinements with DF Lite have to be done outside the wizard.

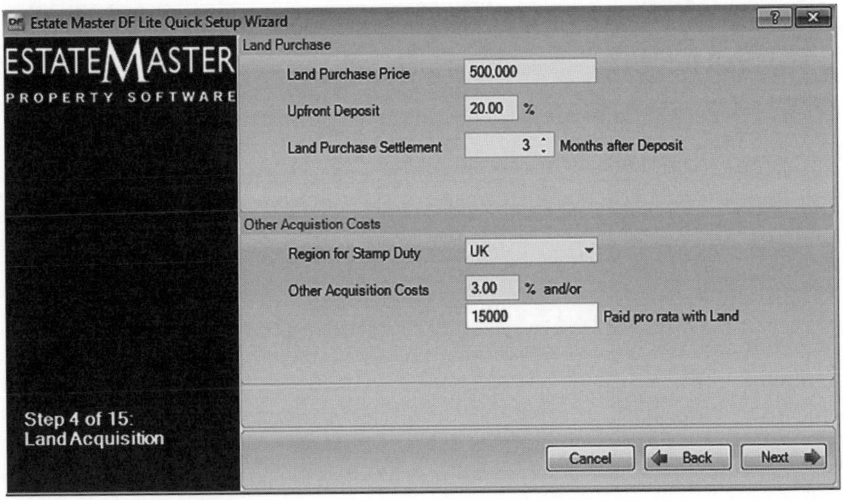

Figure 13.41 Quick set-up screen 4

Screen 8 (Figure 13.45) allows statutory fees such as planning charges etc. to be defined and the date when they are due selected.

The next stage in the set-up wizard deals with land holding costs (Figure 13.46). This includes any rates or council tax liability, any land tax (if appropriate) and any other land holding costs. This is where on-going costs such as site security can be accounted for.

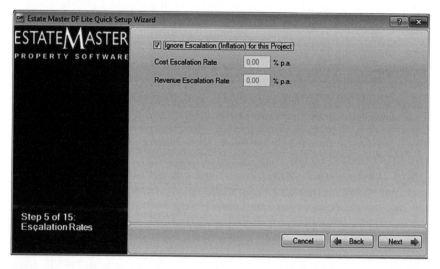

Figure 13.42 Quick set-up screen 5

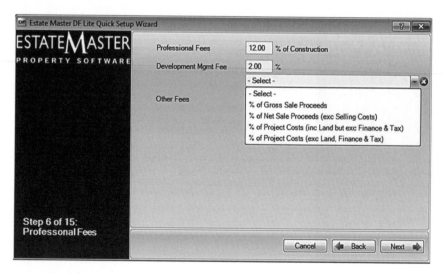

Figure 13.43 Quick set-up screen 6

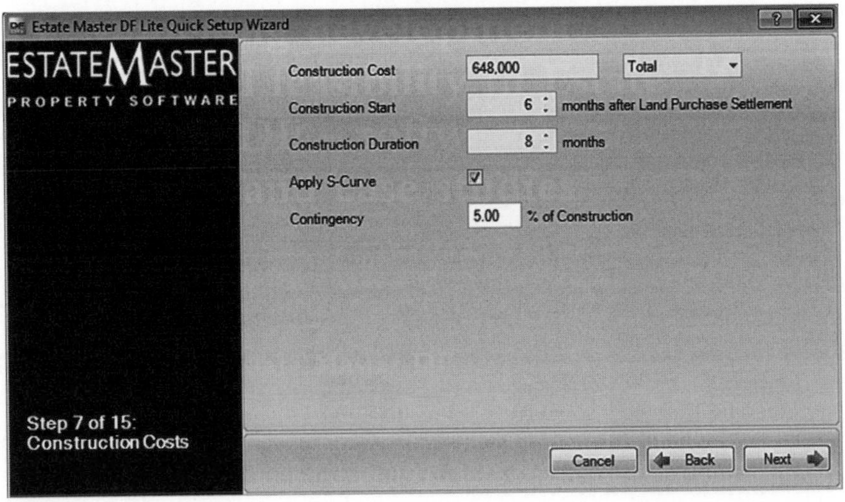

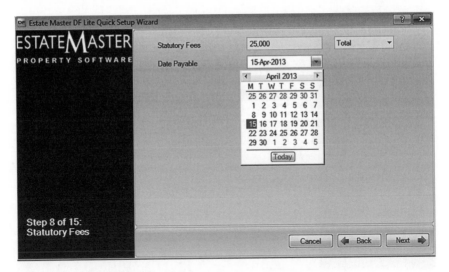

Figure 13.44 Quick set–up screen 7

Figure 13.45 Quick set–up screen 8

The next screen is important as it deals with revenues (Figure 13.47). In terms of residential property this forms a schedule view that allows the definition of numerous types of end product, with values assigned to each and sales assumptions to be made. In this case, the scheme has two different types of residential property, house and apartments. The insert and delete row are used to add and delete property types in this schedule. Again refinements of the entry can be done on completion of the wizard.

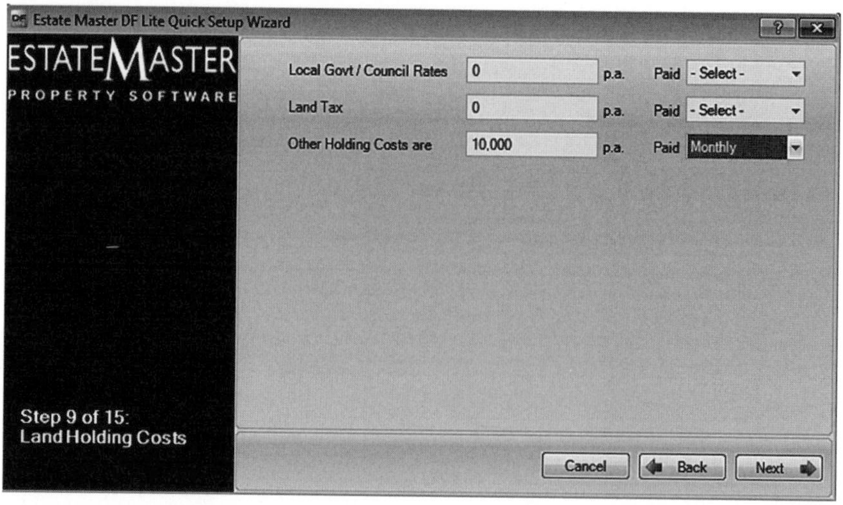

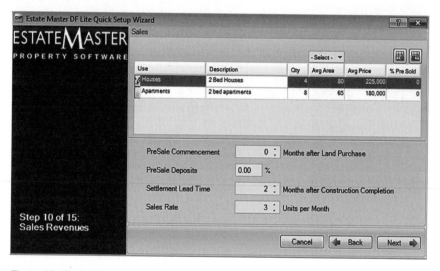

Figure 13.46 Quick set-up screen 9

Figure 13.47 Quick set-up screen 10

The eleventh screen in the wizard deals with any income producing parts of the scheme. On the 'Full' DF program this is also calculated on a separate worksheet. By default on DF Lite the 'ignore' box is checked (Figure 13.48).

Unchecking the box opens up a schedule similar to the residential sales area (Figure 13.49). Here we are assuming that the development has a single retail unit on the ground floor however multiple units are allowed. The appraisal structure is very simple – essentially being confined to annual rent, capitalisation rate and letting up period but is perfectly suitable for a simple appraisal. It is also possible to set up an investment holding period, though leaving it at zero value allows the simplifying assumption of sale coinciding with the letting (just this assumption is usually made in a conventional residual appraisal).

Screen 12 deals with disposal costs – sale and letting fees (see Figure 13.50).

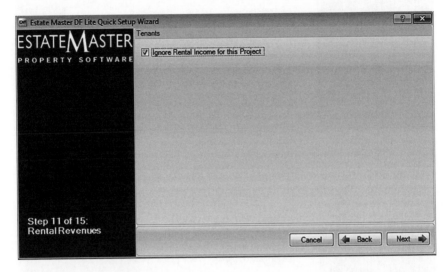

Figure 13.48 Quick set-up screen 11

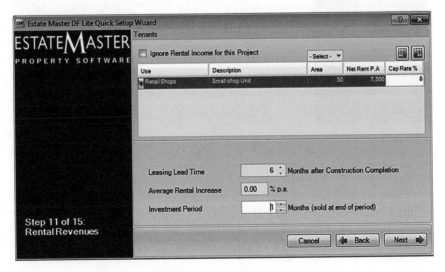

Figure 13.49 Quick set-up screen 11 – completed

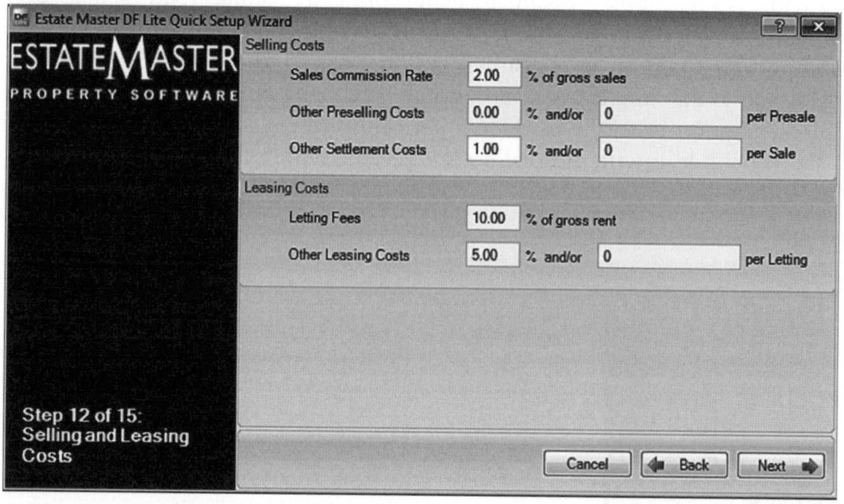

Figure 13.50 Quick set-up screen 12

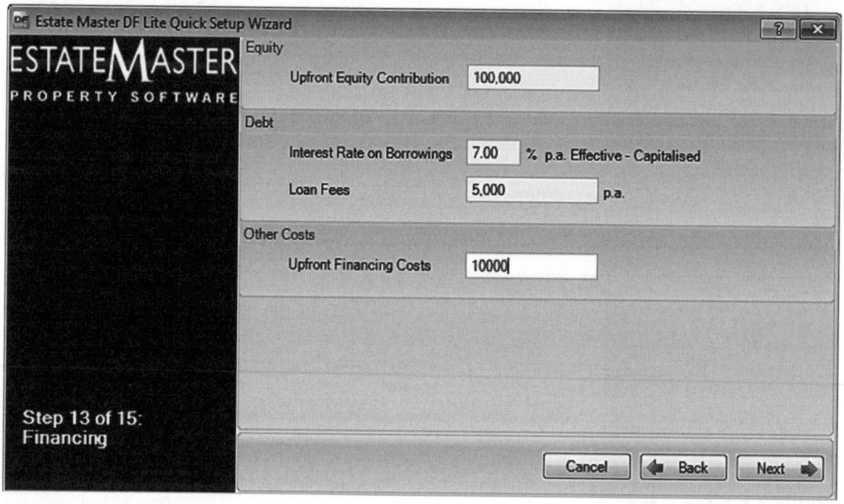

Figure 13.51 Quick set-up screen 13

One major difference between DF Lite and DF 'Full' is that the former has a much simpler finance module (Figure 13.51). With the full DF you can model multiple loans or combinations of loans and equity sharing/partnerships but with the Lite version the user is restricted to upfront equity and a single loan. Again, this is very much in line with the simplifying assumptions made in a simple residual and is perfectly sound for an initial stage appraisal/viability study/land bid calculation.

The penultimate step is to set the target development margins and the target IRR/NPV (Figure 13.52). The target development margin can be defined using the normal industry rules of thumb and is used in the calculation of the land residual (calculated on the summary tab later) whilst the target IRR will define a separate return result (the IRR of the project is independent of this calculation and will be calculated anyway by the software with the results displayed on the summary sheet as well).

That completes the set-up. The user has the opportunity to review all the assumptions made (Figure 13.53) before the set-up is saved.

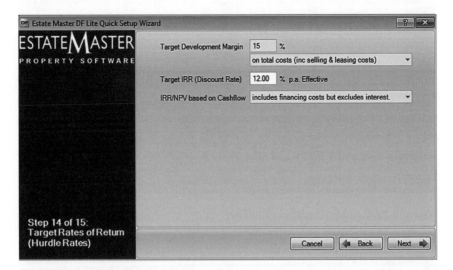

Figure 13.52 Quick set-up screen 14

Figure 13.53 Quick set-up screen 15

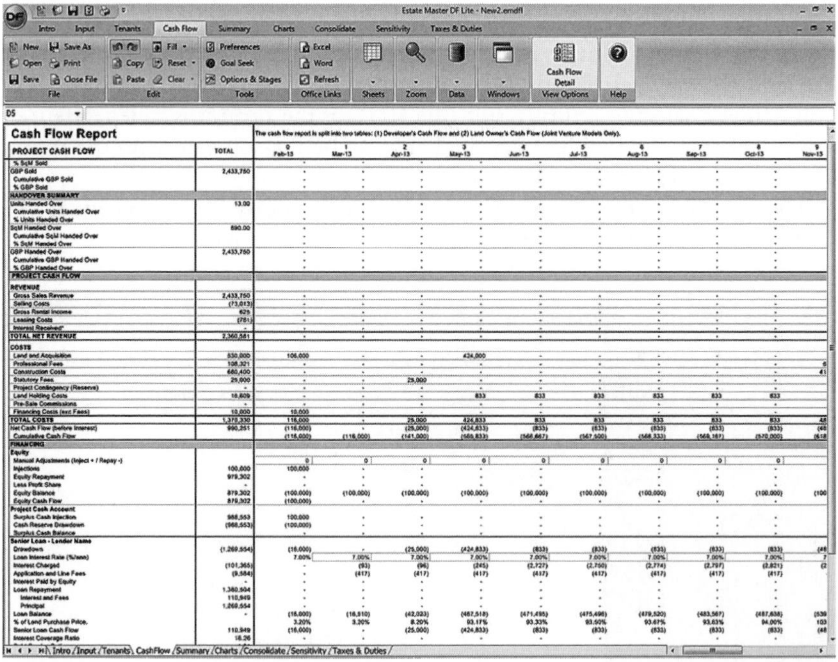

Figure 13.54 Completed project cash flow

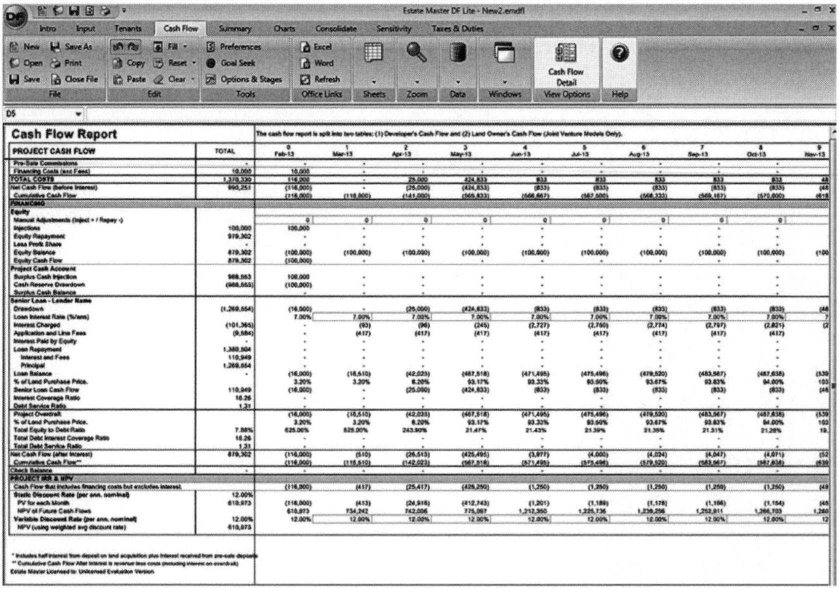

Figure 13.55 Completed financial cash flow

Summary of Project Returns

ESTATEMASTER
PROPERTY SOFTWARE
Development Feasibility

Residential Appraisal
Residential Appraisal
4 Town House plus 8 apartments

Time Span: Feb-13 to Apr-18	Project Size: 12. Units
Type: Residential	1 per 0.08 Ha of Site Area
Status: Under Review	Project Size: 864. GBA
Site Area: 1. Ha	1 per 0.0 Ha of Site Area
FSR: 0:1	

Estate Master Licensed to: Unlicensed Evaluation Version

COSTS & REVENUES					GBP Total	GBP Per Unit	GBP Per Ha of Site Area	GBP Per Total Net Revenue
REVENUE					GBP			
	Quantity	SqM	GBP/Quantity		GBP			
Show **Total Sales Revenue**	13	890.0	187,211.5		2,433,750	202,813	2,433,750	103.1%
Houses	4	320.0	225,000.0		900,000			
Apartments	8	520.0	180,000.0		1,440,000			
Retail Shops	1	50.0	93,750.0		93,750			
Show Less Selling Costs					(73,013)	6,084	73,013	-3.1%
Show Less Purchasers Costs					-	-	-	0.0%
Show NET SALE PROCEEDS					2,360,738	196,728	2,360,738	100.0%
	Average Yield	SqM	GBP/SqM/annum		GBP			
Show Rental Income	8.0%	50.0	150.0		625	52	625	0.0%
Retail Shops	8.0%	50.0	150.0		625			
Show Less Outgoings & Vacancies					-	-	-	0.0%
Show Less Letting Fees					(750)	63	750	0.0%
Show Less Incentives (Rent Free and Fit-out Costs)					-	-	-	0.0%
Show Less Other Leasing Costs					(31)	3	31	0.0%
Show NET RENTAL INCOME					(156)	13	156	0.0%
Show Interest Received					-	-	-	0.0%
TOTAL REVENUE					2,360,581	196,715	2,360,581	100.0%
COSTS								
Show Land Purchase Cost					500,000	41,667	500,000	21.2%
Show Land Transaction Costs					30,000	2,500	30,000	1.3%
Show Construction Costs					680,400	56,700	680,400	28.8%
Show Professional Fees					108,321	9,027	108,321	4.6%
Show Statutory Fees					25,000	2,083	25,000	1.1%
Show Project Contingency (Project Reserve)					-	-	-	0.0%
Show Land Holding Costs					16,609	1,384	16,609	0.7%
Show Pre-Sale Commissions					-	-	-	0.0%
Show Finance Charges (inc. Fees)					19,584	1,632	19,584	0.8%
Show Interest Expense					101,365	8,447	101,365	4.3%
TOTAL COSTS					1,481,279	123,440	1,481,279	62.8%

H ◀ ▶ H \ Intro / Input / Tenants / CashFlow \ Summary / Charts / Consolidate / Sensitivity / Taxes & Duties /

PERFORMANCE INDICATORS		GBP Per Unit	GBP Per Ha of Site Area
1 Net Development Profit		879,302	73,275	879,302
2 Development Margin (Profit/Risk Margin)	Based on total costs (inc selling & leasing costs)	56.54%	
4 Residual Land Value	Based on Target Margin of 15%	973,857	81,155	973,857
5 Net Present Value	Based on Discount Rate of 12% p.a. Nominal	610,973	
6 Benefit Cost Ratio		1.4518	
7 Project Internal Rate of Return (IRR)	Per annum Nominal	44.22%	
8 Residual Land Value	Based on NPV	1,096,413	91,368	1,096,413
Equity IRR	Per annum Nominal	110.07%	
Equity Contribution		100,000	
Peak Debt Exposure		1,368,910	
Equity to Debt Ratio		3.94%	
9 Weighted Average Cost of Capital (WACC)		13.47%	
10 Breakeven Date for Cumulative Cash Flow	Month 23	Jan-2015	
11 Yield on Cost		0.48%	
12 Rent Cover		117 Yrs, 3 Mths	
13 Profit Erosion		N.A.	

Footnotes:
1. Development Profit: is total revenue less total cost including interest paid and received
2. Note: No redistribution of Developer's Gross Profit
3. Development Margin: is profit divided by total costs (inc selling & leasing costs)
4. Residual Land Value: is the maximum purchase price for the land whilst achieving the target development margin.
5. Net Present Value: is the project's cash flow discounted to present value. It includes financing costs but excludes interest.
6. Benefit/Cost Ratio: is the ratio of discounted incomes to discounted costs and includes financing costs but excludes interest.
7. Internal Rate of Return: is the discount rate where the NPV above equals Zero.
8. Residual Land Value (based on NPV): is the purchase price for the land to achieve a zero NPV.
9. The Weighted Average Cost of Capital (WACC) is the rate that a company is expected to pay to finance its assets.
10. Breakeven date for Cumulative Cash Flow: is the last date when total debt and equity is repaid (ie when profit is realised).
11. Yield on Cost is Current Net Annual Rent divided by Total Costs, including all Selling Costs.
12. The total net development profit divided by the current net annual rental expressed as a number of years/months.
13. The period of time post practical completion that it can remain unsold (but leased out) until finance and land holding costs erodes the profit for the development to zero.

RETURNS ON FUNDS INVESTED	Equity	Senior Loan
		Lender None
1 Funds Invested (Cash Outlay)	100,000	1,269,554
% of Total Funds Invested	3.79%	48.11%
2 Peak Exposure	100,000	1,368,910
Date of Peak Exposure	Feb-13	Nov-14
Month of Peak Exposure	Month 0	Month 21
Weighted Average Interest Rate	N.A.	7.00%
Interest Charged	-	101,365
Line Fees Charged	-	9,584
Application Fees Charged	-	-
3 Total Profit to Funders	879,302	110,949
4 Margin on Funds Invested	879.30%	8.74%
5 Payback Date	Apr-15	Jan-15
Month of Payback	Month 26	Month 23
6 IRR on Funds Invested	110.07%	7.68%
7 Equity to Debt Ratio		7.88%
8 Loan to Value Ratio	4.11%	56.25%
9 Loan Ratio	20.00%	276.10%
	of Land Purchase Price.	of Land Purchase Price.

Footnotes:
1. The total amount of funding injected into the project cash flow.
2. The maximum cash flow exposure of that equity/debt facility including capitalised interest.
3. The total repayments less funds invested, including profit share paid or received.
4. Margin is net profit divided by total funds invested (cash outlay).
5. Payback Date for the equity/debt facility is the last date when total equity/debt is repaid.
6. IRR on Funds Invested is the IRR of the equity cash flow including the return of equity and realisation of project profits.
7. Equity to Debt Ratio is the amount of equity contributed into the project as a percentage of debt funding.
8. Loan to Value ratio is the Peak Equity/Debt Exposure divided by Total Sales Revenue.
9. Loan Ratio is the total funds invested by the lender (cash outlay) divided by the nominated ratio calculation method. It includes capitalised interest and fees.

Figure 13.56 Completed project summary sheet

Sensitivity Analysis

ESTATE**MASTER**	Residential Appraisal
PROPERTY SOFTWARE	Residential Appraisal
Development Feasibility	4 Town House plus 8 apartments
	Estate Master Licensed to: Unlicensed Evaluation Version

SENSITIVITY TABLE	Change %	Net Dev. Profit	NPV	Dev. Margin	Project IRR	Equity IRR
Base Case (No Variation)	0.0%	879,302	610,973	56.54%	44.22%	110.07%
Land Acquisition Costs	-5.0%	909,694	637,277	59.66%	46.19%	111.61%
	-3.0%	897,548	626,764	58.40%	45.40%	111.00%
	3.0%	861,025	595,154	54.73%	43.07%	109.12%
	5.0%	848,823	584,593	53.53%	42.32%	108.47%
Construction Costs	-10.0%	961,671	679,255	65.30%	48.06%	114.14%
	-5.0%	920,486	645,114	60.80%	46.14%	112.14%
	5.0%	838,117	576,832	52.51%	42.32%	107.90%
	10.0%	796,933	542,691	48.67%	40.43%	105.65%
Construction Period *	-20.0%	910,902	684,413	59.79%	57.77%	139.75%
	-10.0%	895,247	647,737	58.17%	50.21%	126.08%
	10.0%	853,972	556,765	54.04%	37.34%	97.05%
	20.0%	837,437	522,354	52.44%	33.89%	89.84%
End Sale Values	-5.0%	761,002	517,012	49.04%	40.07%	103.60%
	-3.0%	808,322	554,596	52.05%	41.76%	106.28%
	3.0%	950,282	667,349	61.03%	46.61%	113.60%
	5.0%	997,602	704,933	64.01%	48.16%	115.83%
Capitalisation Rate	-0.5%	885,358	615,648	56.93%	44.41%	110.38%
	-0.2%	881,631	612,771	56.69%	44.29%	110.19%
	0.2%	877,086	609,262	56.40%	44.16%	109.95%
	0.5%	873,959	606,847	56.21%	44.06%	109.79%
Sales Span **	-30.0%	882,903	620,623	56.91%	45.69%	110.25%
	-20.0%	880,912	616,692	56.71%	45.10%	110.15%
	20.0%	878,371	605,291	56.45%	43.39%	110.02%
	30.0%	876,349	601,388	56.25%	42.85%	109.91%
Rental Levels	-20.0%	861,166	596,971	55.40%	43.67%	109.13%
	-10.0%	870,234	603,972	55.97%	43.95%	109.60%
	10.0%	888,370	617,973	57.11%	44.50%	110.53%
	20.0%	897,438	624,974	57.68%	44.78%	110.99%
Debt Interest Rates	-2.0%	909,203	610,973	59.61%	44.22%	111.58%
	-1.0%	894,349	610,973	58.07%	44.22%	110.84%
	1.0%	864,060	610,973	55.02%	44.22%	109.28%
	3.0%	832,979	610,973	52.02%	44.22%	107.63%
Discount Rate	18.0%		460,560			
	19.0%		437,429			
	20.0%		414,820			
	21.0%		392,721			

* Variation to Construction Period in sensitivity table delays span dates for Construction and start and span for Professional Fees, Statutory Contributions and Misc. Costs. Delays start but not span date for Sales, Rental and Other Income. Delays the span for Land Holding costs. Has no effect on Land Purchase or Finance costs (except interest).
** Varies span date for Pre-Sale Exchange and Settlement periods, but not commencement dates.

Figure 13.57 Completed project sensitivity analysis

Once the assumptions made in the set-up have been saved the program returns you to the main screen view but now all of the data that has been entered have been automatically put into the DF framework. Similarly, the data has been entered into both the Input and the Tenants worksheets. As has been mentioned several times before, this is fully editable just as it would be if the data had been entered into the system in the traditional format. This allows adjustment and refinement of the project as well as creating a file that can be translated into a DM file for project monitoring at a later date.

The software produces both a project (Figure 13.54) and finance (Figure 13.55) cash flow.

The programme produces the same informative performance indicators as the full program (Figure 13.56 – note that the user is prompted to calculate the land residual when opening this worksheet (previous page)) and also the powerful sensitivity analysis module (Figure 13.57).

Conclusion

DF Lite offers up a much simpler and quicker way of using the power and reliability of the Estate Master software suite. It has huge advantages over self-created spreadsheets (accuracy, reliability, professional presentation, speed of construction) at relatively low year-on-year cost. It also provides a solution that integrates with tools that can be used in development project management.

14 Conclusions

One of the criticisms I have frequently heard voiced about software packages such as Argus Developer and Estate Master DF is that they make appraisal too easy, or rather they trap the unwary or inexperienced into rushing into calculation and, therefore, unwise and half-baked conclusions.

I would stress that despite the concentration this book has on the use of Excel, Argus Developer and Estate Master DF, development appraisal is a process that should not be rushed. The calculations, effectively the use of the models described here, should be just the tip of the iceberg in the process. For every five minutes spent using the models perhaps an hour or more should be spent collecting, checking and validating the data used within the models.

But this does not invalidate the use of the software tools, far from it. The job of the valuer, development appraiser and developer is difficult enough; all of us who fulfil these roles need as much assistance as possible. The software explored on these pages are only tools, and like any tool, if they are used improperly then any failure is not down to them but the operator. What they do achieve is to make the job much easier – they provide a reliable, consistent and easy to use service to the valuer/appraiser/developer. They also give these parties the ability to extend their skills and abilities in analysis – particularly in exploring complex financial arrangements and in exploring risk.

They are an invaluable part of the armoury of anyone conducting development appraisals and their importance will only grow in the future.

Appendix A – The Town and Country Planning (Use Classes) Order

The Town and Country Planning (Use Classes) Order 1987 is as follows:

SCHEDULE

PART A

Class A1. Shops

Use for all or any of the following purposes –

(a) for the retail sale of goods other than hot food,
(b) as a post office,
(c) for the sale of tickets or as a travel agency,
(d) for the sale of sandwiches or other cold food for consumption off the premises,
(e) for hairdressing,
(f) for the direction of funerals,
(g) for the display of goods for sale,
(h) for the hiring out of domestic or personal goods or articles,
(i) for the reception of goods to be washed, cleaned or repaired, where the sale, display or service is to visiting members of the public.

Class A2. Financial and professional services

Use for the provision of –

(a) financial services, or
(b) professional services (other than health or medical services), or
(c) any other services (including use as a betting office) which it is appropriate to provide in a shopping area, where the services are provided principally to visiting members of the public.

Class A3. Food and drink

Use for the sale of food or drink for consumption on the premises or of hot food for consumption off the premises.

PART B

Class B1. Business

Use for all or any of the following purposes –

(a) as an office other than a use within class A2 (financial and professional services),
(b) for research and development of products or processes, or
(c) for any industrial process, being a use which can be carried out in any residential area without detriment to the amenity of that area by reason of noise, vibration, smell, fumes, smoke, soot, ash, dust or grit.

Class B2. General industrial

Use for the carrying on of an industrial process other than one falling within class B1 above or within classes B3 to B7 below.

Class B3. Special Industrial Group A

Use for any work registrable under the Alkali, etc. Works Regulation Act 1906(1)(a) and which is not included in any of classes B4 to B7 below.

Class B4. Special Industrial Group B

Use for any of the following processes, except where the process is ancillary to the getting, dressing or treatment of minerals and is carried on in or adjacent to a quarry or mine –

(a) smelting, calcining, sintering or reducing ores, minerals, concentrates or mattes;
(b) converting, refining, re-heating, annealing, hardening, melting, carburising, forging or casting metals or alloys other than pressure die-casting;
(c) recovering metal from scrap or drosses or ashes;
(d) galvanizing;
(e) pickling or treating metal in acid;
(f) chromium plating.

Class B5. Special Industrial Group C

Use for any of the following processes, except where the process is ancillary to the getting, dressing or treatment of minerals and is carried on in or adjacent to a quarry or mine –

(a) burning bricks or pipes;
(b) burning lime or dolomite;
(c) producing zinc oxide, cement or alumina;
(d) foaming, crushing, screening or heating minerals or slag;
(e) processing pulverized fuel ash by heat;
(f) producing carbonate of lime or hydrated lime;
(g) producing inorganic pigments by calcining, roasting or grinding.

Class B6. Special Industrial Group D

Use for any of the following processes –

(a) distilling, refining or blending oils (other than petroleum or petroleum products);
(b) producing or using cellulose or using other pressure sprayed metal finishes (other than in vehicle repair workshops in connection with minor repairs, or the application of plastic powder by the use of fluidised bed and electrostatic spray techniques);
(c) boiling linseed oil or running gum;
(d) processes involving the use of hot pitch or bitumen (except the use of bitumen in the manufacture of roofing felt at temperatures not exceeding 220°C and also the manufacture of coated roadstone);
(e) stoving enamelled ware;
(f) producing aliphatic esters of the lower fatty acids, butyric acid, caramel, hexamine, iodoform, napthols, resin products (excluding plastic moulding or extrusion operations and producing plastic sheets, rods, tubes, filaments, fibres or optical components produced by casting, calendering, moulding, shaping or extrusion), salicylic acid or sulphonated organic compounds;
(g) producing rubber from scrap;
(h) chemical processes in which chlorphenols or chlorcresols are used as intermediates;
(i) manufacturing acetylene from calcium carbide;
(j) manufacturing, recovering or using pyridine or picolines, any methyl or ethyl amine or acrylates.

Class B7. Special Industrial Group E

Use for carrying on any of the following industries, businesses or trades –

Boiling blood, chitterlings, nettlings or soap.
Boiling, burning, grinding or steaming bones.
Boiling or cleaning tripe.
Breeding maggots from putrescible animal matter.
Cleaning, adapting or treating animal hair.
Curing fish.
Dealing in rags and bones (including receiving, storing, sorting or manipulating rags in, or likely to become in, an offensive condition, or any bones, rabbit skins, fat or putrescible animal products of a similar nature).
Dressing or scraping fish skins.
Drying skins.
Making manure from bones, fish, offal, blood, spent hops, beans or other putrescible animal or vegetable matter.
Making or scraping guts.
Manufacturing animal charcoal, blood albumen, candles, catgut, glue, fish oil, size or feeding stuff for animals or poultry from meat, fish, blood, bone, feathers, fat or animal offal either in an offensive condition or subjected to any process causing noxious or injurious effluvia.
Melting, refining or extracting fat or tallow.
Preparing skins for working.

Class B8. Storage or distribution

Use for storage or as a distribution centre.

PART C

Class C1. Hotels and hostels

Use as a hotel, boarding or guest house or as a hostel where, in each case, no significant element of care is provided.

Class C2. Residential institutions

Use for the provision of residential accommodation and care to people in need of care (other than a use within class C3 (dwelling houses)).
Use as a hospital or nursing home.
Use as a residential school, college or training centre.

Class C3. Dwellinghouses

Use as a dwellinghouse (whether or not as a sole or main residence) –

(a) by a single person or by people living together as a family, or
(b) by not more than 6 residents living together as a single household (including a household where care is provided for residents).

PART D

Class D1. Non-residential institutions

Any use not including a residential use –

(a) for the provision of any medical or health services except the use of premises attached to the residence of the consultant or practioner,
(b) as a crèche, day nursery or day centre,
(c) for the provision of education,
(d) for the display of works of art (otherwise than for sale or hire),
(e) as a museum,
(f) as a public library or public reading room,
(g) as a public hall or exhibition hall,
(h) for, or in connection with, public worship or religious instruction.

Class D2. Assembly and leisure

Use as –

(a) a cinema,
(b) a concert hall,
(c) a bingo hall or casino,
(d) a dance hall,
(e) a swimming bath, skating rink, gymnasium or area for other indoor or outdoor sports or recreations, not involving motorised vehicles or firearms.

Source: http://www.legislation.gov.uk/uksi/1987/764/schedule/made (accessed February 2013)

Index

Abu Dhabi 25
access 37, 42–3
accumulative cash flows 69–76
acquisition *see* land
advertising 67 *see also* marketing
Affordable Homes Programme 102
All Risks Yield (ARY) 68
appraisal 90; basic equation 17–21;
    theory of 12–17
archaeology 39
architects 50, 188
area screens 135–40; multi-building
    project 154–5; operated assets 191;
    residential 199, 201, 205, 207–9,
    211
Argus Developer 110–21, 260;
    commercial 121–49; interest sets
    178–84; mixed-use 160–7; multiple
    buildings/types 152–9; operated
    assets 185–96; phasings 172–8;
    profitability calculation 149–52;
    residential (complex) 204–13;
    residential (simple) 196–204;
    sensitivity analysis 81, 213–18;
    timings 167–71
asbestos 40
auditing 106
Australia 29, 219–20

Bank of England 31, 44–5
banks 31
Barker, K. 52
base rate 31, 45
Bath 25–6, 38
BCIS *see* Building Cost Information
    Service
benchmarks 22–4
Boots 8
boundaries 27
break clauses 60

British Land 7
Brown, P. 97, 100–1
building contracts 15–16, 27
Building Cost Information Service
    (BCIS) 50, 136
building dimensions 45
Burj Khalifa 25

Cadogan Estates 7
capital value 58–60, 64–5, 125
capital value *see* also yield
carbon footprint 29
care homes 185
cash flow 62, 68–79, 94; Developer
    116–18, 132, 144–6; Estate Master
    228–9; Estate Master Lite 256, 258;
    multiple phases 174–7; operated assets
    192–3; residential 202–3, 210–11
central banks 44–5
Christmas 29
Circle Developer 110
city centres 25–6, 28, 37, 39
City of Culture 29
commercial development 7, 13, 24,
    42–4; Developer 121–49; Estate
    Master 225; estimating values 54–8;
    mixed-use 160–7; multi-building
    152–9; profitability (Developer)
    149–52; profitability (Estate Master)
    229–39; specialised 60–1
commercial lenders 12–13
Community Infrastructure Levy (CIL)
    52, 102
companies 6–9
complexity 10, 59, 70, 94, 97, 107;
    Developer 152
compulsory purchase 35
condition 57
conservation areas 26, 38
constraints 25–34

construction 34, 46, 48–9; Developer 136–7, 142–3, 207–8; Estate Master 222, 225, 236; Estate Master Lite 249; residual approach 65; sensitivity analysis 215–17
consultants 15, 42, 51
contaminated land 29, 40–1
contingency 65
contractors 34
contractual issues 26–8
contributions 51, 102
Coppergate 40
costs 15, 48–54, 222–5 *see also* construction; letting; professional fees; sales
credit rate 178, 199

data collection 36–41, 136; estimating costs 48–54; estimating values 54–61; market analysis 41–5
debt finance 30–1
definition tab 113–16, 139, 141–4; multi-building 153, 156; multi-phase 174–5; residential 201–2, 205
definitions 5
demand and supply 43–4, 58
demolition 48, 113, 115–16
depreciation 60
design and build 15, 34, 51
detailed appraisal 14–16
Developer *see* Argus Developer
Development Appraisal Tool (DAT) 98, 101–7
discounting 18, 20–1, 64, 69–74, 76–8 *see also* interest
disposal costs *see* letting; sales
distribution 164, 169
Dubai 25

easements 28
economic trends 44–5
Edinburgh 25–6, 38
Eire 44
elasticity 58
elemental cost approach 49
energy costs 10, 29
engineers 50
envelope 25–34
environmental constraints 28–9
Environmental Protection Act 40
equations 17–21
equity funds 30, 34n1, 65–6, 234, 239; opportunity cost 52
errors 1, 70, 94–5, 97–8, 106–8

escalation 222, 229
Estate Master DF 106, 219–29, 260; Lite 246–59; profitability calculation 229–39; sensitivity analysis 81–2, 239–46
Estates Gazette 42
estimation: of costs 48–54; of values 54–61
Europe 29, 39
Excel *see* spreadsheets
expenditure assumptions 123–4
export options 120

fashion 10
Federal Reserve 44
final appraisals 14–16
finance 9, 12–13, 30–3; cash flow 144, 146, 178; Developer 125, 137–8; Estate Master 231, 234, 238; Estate Master Lite 254; interest sets 178–84; objectives 22–4; opportunity costs 52; residual approach 65–6; software use in sector 93–4; trends 44–5
flood risk 28–9
flora and fauna 39
freehold 53, 114, 139–40, 165

gap funding 32–3
gearing 19, 21, 29–30, 108
geology 39
global financial crisis 31, 44–5
golf courses 61, 185–96
Google Earth 37
grants 32–3, 102
graphical analysis 120
Greece 44
Gross Development Value (GDV) 31
Grosvenor 7

health and safety 50, 53
heritage 39–40
historic areas 26
Homes and Communities Agency (HCA) 98, 101–7
hotels 29, 61, 185, 188
housing associations 6, 8, 98

inception date 46
industrial development 42, 172–3
inflation 45
infrastructure projects 14, 22
initial appraisal 14–16
input worksheet 232
institutional lease 55

insurance 29
interest rates 20–1, 28, 31–2, 34n1, 45, 47; Developer 125, 127–9; multiple sets 178–84; opportunity cost 52; residential 199; residual approach 65–6
Internal Rate of Return (IRR) 24, 76, 78, 129; Estate Master 244–5; Estate Master Lite 255
internationalisation 10
internet 42–3
ITZA 161–2

Japanese Knotweed 39
joint ventures 32–3, 220
Jorvik Centre 40

Katz, A. 110

Lake District 38
land acquisition 15, 27; Estate Master 220–2, 232–4; Estate Master Lite 249
land assembly 35–6
land banks 9, 35
land holding costs 52–3, 250
Land Securities 7
land use *see* planning
land value 17; Developer 129; multi-phase 172, 177–8; residential 197–204
landscaping 141–2
legal fees 53–4
lenders 12–13, 30–3, 45
letting 47–8; Developer 139, 170–1; fees 53, 66, 143, 225, 253; lease terms 55–6, 59–60; supply/demand 43–4
Libre Office 94
life assurance 8
limits 25–34
loans *see* finance
local authority 35–7, 40, 102, 225
Local Development Framework 36
Localism Act 52
location 37–8, 57–8, 60
London 57

major events 29
malls 35, 42
managers 50–1
Manchester 37
margins *see* profitability; returns; yield

marinas 185
market 6, 9–11, 58; analysis 41–5; timing 29–30, 45–8
Market Rental Value (MRV) 137–8
marketing 67; Developer 143–4; Estate Master 234, 237
membership fees 189
mezzanine loans 32–3, 178
mixed-use: Developer 160–7; Estate Master Lite 246–59
Monte Carlo analysis 82, 244
MS Excel *see* spreadsheets
multi-building projects 152–9; mixed-use 160–7

National Freehold 7
National Parks 38
Net Present Value (NPV) 70, 78, 255
netting down 54
New Zealand 29, 219
non-profits 6, 14, 98

objectives 22–4
offices 23, 42–3, 57
Olympics 29
online retail 42
Open Office 94
operated assets 185–96
opportunity costs 52, 65–6, 234
Oxford 25–6

parking 136
performance indicators 230, 239, 242–3, 258
phasing 46–7; Developer 112–13, 132–4, 172–8; operated assets 185–6; residential 204–6; residual approach 63–4
physical package 56–7, 60
planning 26, 46, 102; contributions 51–2; data collection 36–7; Developer 143; Estate Master 225
Planning Gain Supplement (PGS) 52
Planning Policy Guidelines (PPGs) 39
Planning Policy Statements (PPS) 36, 39–40
political constraints 28
pollution 29, 40
Portman 7
preconstruction period 46, 72

preference screen 220–1
present value *see* discounting
prices 9, 27, 31, 45
prime locations 37
Private Finance Initiative (PFI) 23, 34, 51
probabilities 82, 85, 89; Estate Master 239, 244
procurement 33–4
professional fees 50–1; Developer 123; Estate Master 222, 234–6; Estate Master Lite 249; residual approach 65
profitability 13, 17, 24, 67; calculating in Developer 149–52; calculating in Estate Master 229–39; sensitivity analysis 81–2
project management 16, 50, 201
project objectives 22–4
project tab 111–13
property markets 6, 9–11
Property Week 42
proprietary systems 1, 16, 61–2, 91, 260; advantages over Excel 93–4, 108–9; criticism 21 *see also* Argus Developer; Estate Master DF
Prudential 8
public funding 32–3
Public-Private Partnerships (PPPs) 23, 34, 51

quantity surveyors 50

Real Estate Investment Trusts (REITs) 7, 30
receipts 124–5
rent free period 125, 138
rental value 37–8, 55–8; Developer 137–8, 160; residual approach 64; sensitivity analysis 81–2, 85, 215–17
reporting module 120
residential development 6, 9–10, 35, 42; benchmark return 13, 24; complex 204–13; estimating values 54; mixed-use (Developer) 160–7; mixed-use (Estate Master) 249–59; simple 196–204
residual approach 10, 18–19, 62–9, 78, 204; cash flow models 69–76
resort locations 29, 61
retail 35, 37, 42, 56–8, 160–1
returns: on cost 24, 129, 131; on funds invested 231, 234, 243 *see also* yield

RICS 136
rights over the land 28
risk 13, 21, 80; residential vs. commercial 197–8 *see also* sensitivity analysis; uncertainty

S106 143
safety 50, 53
sales: Developer 123, 143, 163–5; Estate Master 225; Estate Master Lite 253; residential 208–9, 213; residual approach 66–7
saving data 122–3
scenarios 81–2, 85, 88–9; Developer 214–15; Estate Master 239, 244
Scotland 52
Section-75 52
Section-106 27, 225
security 53
selling costs 53
senior debt 32
sensitivity analysis 30, 80–9; Developer 121, 213–18; Estate Master 229, 239–46; Estate Master Lite 258
servicing 37
shape 57
shops *see* retail
short term 14
simulation 82, 244–6
site *see* land; location
size 56
social housing 6, 8, 27, 98
social objectives 23–4
software *see* proprietary systems
solicitors 53
South East 172
Spain 44
specification 57
sporting events 29
spreadsheets 1, 16, 93–101, 108–9, 192; cash flows 68–70, 77; Estate Master 219, 229; HCA DAT 102–6; sensitivity analysis 81–2, 85, 88–9; typical errors 106–8
stages 46–7 *see also* phasing
Stamp Duty Land Tax 53–4
Standard Life 8
statutory fees 225–6, 249–50
stock market 31
summary tab 118–19
superficial error approach 49
supermarkets 42
supply and demand 43–4, 58
surveyors 9, 50

targets 22–4
taxation 53; Estate Master 220; Estate
    Master Lite 248, 250; VAT 137–8,
    146
tenancy: schedule 225, 227, 238, 240;
    status/size of tenant 58–9 *see also*
    letting
theory 12–17
timescale 97; Developer 112–13, 132–4,
    137, 167–71; mixed-use 160, 162–5;
    multi-phase 172–3; operated assets
    185–6; residential 201, 205–6; residual
    approach 63–4 *see also* market
    timing
title 27 *see also* freehold
topography 38
Town and Country Planning Act 5, 52,
    261–5
transport 10, 42–3
trees 39
trends 41–5
turnkey 34, 51

UAE 219
UK 26, 33, 55
uncertainty 24, 54 *see also* risk; sensitivity
    analysis

unit sales 161, 163
Universal Business Rate (UBR) 53
urban regeneration 10
USA 29, 44, 186

validity of inputs 21
value 44; estimating 54–61; residual
    approach 64–5 *see also* land value
variable contract 27
variations 12–13
VAT 137–8, 146
vegetation 39
viability studies 14–15
Visual Developer 110
volatility 108
volume builders 6, 9, 49

Wall Street 93
warehouses 172–3

yield 24, 58–60; Developer 139; residual
    approach 64; sensitivity analysis 81,
    85, 215–18; spreadsheets 68 *see also*
    returns
York 40

Zone A 161